D1236664

Molecular Shapes

MOLECULAR SHAPES

Theoretical Models of Inorganic Stereochemistry

JEREMY K. BURDETT

Department of Chemistry
The University of Chicago

A WILEY-INTERSCIENCE PUBLICATION

JOHN WILEY & SONS,
New York ● Chichester ● Brisbane ● Toronto

Library of Congress Cataloging in Publication Data:

Burdett, Jeremy K   1947-
  Molecular shapes.

  "A Wiley-Interscience publication."
  Includes index.
  1. Molecular structure. 2. Stereochemistry.
I. Title.

QD461.B97   541.2'23   80-15463
ISBN 0-471-07860-3

Printed in the United States of America

10 9 8 7 6 5 4 3 2 1

Preface

Recent years have seen an explosion both in the number of new structural types that have been synthesized and in the number of structural studies undertaken by X-ray and neutron diffraction methods. Theoretical ideas to rationalize these new and exciting structural results are much in demand. The theoretical world is however at present divided as to how to meet this need. One group prefers to perform high-quality molecular orbital calculations on a limited series of molecular configurations and relies heavily on computer-generated numerical data. The other group uses molecular orbital results from much cruder molecular orbital methods, leans to a much smaller extent on exact numerology, and tries to find a symmetry or overlap explanation whenever possible. This second group has been led by the sterling efforts of Roald Hoffmann whose analogous treatment of the organic field a decade ago provided the impetus for a revolution in that area which brought theory and experiment much closer together.

Both approaches have their role to play. Good calculations are necessary to be able to reproduce detailed electronic properties of molecules such as photoelectron spectra, for example. But simple theories are of more interest to the average chemist. As L. S. Bartell has written, "By now chemists are fully accustomed to the gloomy rule of thumb that the more exact the quantum molecular calculation the more obscured by complexity is the physical interpretation and the less intelligible are the guidelines afforded to nonspecialists."[10] Similarly, in summarizing the inorganic chemist's view of molecular theories in 1970, J. W. Faller concluded: "It is the simple arguments, some naive, some well-founded on symmetry considerations, all drastic oversimplifications, that are most successful and practically useful" (p. 415 in Ref. 194).

Our aim in this book is to provide the reader with an account of various simple theoretical models of inorganic stereochemistry. There is no monopoly on ways to explain the same structural result, but some of the models are more generally applicable than others. The angular overlap model, for example, is widely used throughout this volume and provides a theoretical link between main group and transition metal stereochemistry, two areas that are often viewed separately. Our goal is to lead to an understanding of molecular structure rather than a cataloguing of numerical structural results in the sense of Hoffmann's comment that ". . . to understand an observable means being able to predict, albeit qualitatively, the result that a perfectly reliable calculation would yield for that observable."[93] However it should be constantly borne in mind when reading and using this book that our simple models are often drastic simplifications of the quantum mechanical "truth." In one sense they are just ways to collect together and organize experimental observations.

We hope to show in a general way how a small theoretical armory may be employed to tackle a range of structural problems. However the book is not a collection of explanations for each unusual structural quirk. (It could well be subtitled "The Structures of Simple Molecules.") The references similarly are not all-embracing. The majority of structural results quoted are to be found in Wells' elegant volume,[218] in Pearson's book,[163] or in the extremely valuable structural bibliographies provided by BID-ICS[19] and the Chemical Society's Specialist Periodical Reports.[193] A comprehensive set of references covering experimental and theoretical studies on structural aspects of small main group molecules is to be found in Gimarc's recent book.[73]

One of the problems facing a writer of a book of this type is whether to include a discussion of the use of group theoretical methods in constructing molecular orbitals. This is done so well in Cotton's[40] little book that the motivation for including a substantial chapter in this volume was small. Although no significant discussion is included on this topic, the introductory chapters do include a collection of material basic to the study of molecular orbital theory.

The book has been read by several people at various stages of its gestation period. Particular thanks are due to T. A. Albright, R. S. Berry, and R. L. DeKock for their comments and criticisms, to Peri Gruber who drew the figures, and to Nancy Trombetta who typed the manuscript and made the many revisions and corrections with patience and good humor. To Professor Albright I am particularly indebted for his permission to reproduce the diagrams of Figure 13.8.

JEREMY K. BURDETT

Chicago, Illinois
May 1980

Contents

A Note on Nomenclature

The following symbols and abbreviations appear in the text and are defined here for convenience:

A = main group atom

M = transition metal atom

H = hydride ligand (with σ orbital only)

Y = ligand in general (with π and σ orbitals)

X = halogen

L = ligand with π acceptor properties

B = shared electron pair on VSEPR scheme

E = unshared electron pair

HOMO = highest occupied molecular orbital

LUMO = lowest unoccupied molecular orbital

ls = low spin

is = intermediate spin

hs = high spin

ϕ = atomic orbital

ψ = molecular orbital

Ψ = wavefunction describing electronic state

ϵ = interaction energy

$\Sigma(\sigma)$ = σ stabilization energy

$\Sigma(\pi)$ = π stabilization energy

Molecular Shapes

1 Introductory

Several simple theoretical results are described in this chapter that facilitate our understanding of the molecular orbital structures of molecules. In addition to these, we make extensive use of symmetry and group theoretical arguments, a powerful way to derive wavefunctions and classify energy levels in molecules.

1.1 Perturbation Theory

Of fundamental importance in the application of molecular orbital ideas to chemical problems is the use of perturbation theory to derive the wavefunctions and energies of a complex system from those of a much simpler one, where such information may be more readily obtained. We shall use the theory in many places throughout this book to view, for example, transition metal—or main group—ligand interaction or to understand how molecular orbitals mix when a molecule is distorted.

Consider a system described by a Hamiltonian $\mathcal{H}^{(0)}$ with a set of eigenvalues $E_m^{(0)}$ and eigenfunctions $\Psi_m^{(0)}$ such that

$$\mathcal{H}^{(0)}\Psi_m^{(0)} = E_m^{(0)}\Psi_m^{(0)} \tag{1.1}$$

Let us now disturb the system by applying a perturbation such that the perturbed system is described by the new Hamiltonian $\mathcal{H} = \mathcal{H}^{(0)} + \lambda\mathcal{H}'$ where λ is a parameter that allows us to gradually switch on the perturbation. Generally we may write for the new energies and wavefunctions

$$E' = E^{(0)} + \lambda E^{(1)} + \lambda^2 E^{(2)} + \cdots \tag{1.2}$$

$$\Psi' = \Psi^{(0)} + \lambda\Psi^{(1)} + \lambda^2\Psi^{(2)} + \cdots$$

where the $E^{(n)}$ and $\Psi^{(n)}$ are the nth order corrections to the energy and wavefunction, respectively. For the perturbed system the Schrödinger equation becomes for the mth state

$$[\mathcal{H}^{(0)} + \lambda\mathcal{H}' - (E^{(0)} + \lambda E_m^{(1)} + \lambda^2 E_m^{(2)}$$
$$+ \cdots)][\Psi_m^{(0)} + \lambda\Psi_m^{(1)} + \lambda^2\Psi_m^{(2)}] = 0 \quad (1.3)$$

If this is valid for all values of λ, then the functions multiplying each power of λ must each individually be zero in Equation 1.3. In first order (terms in λ),

$$(\mathcal{H}^{(0)} - E_m^{(0)})\Psi_m^{(1)} + (\mathcal{H}' - E_m^{(1)})\Psi_m^{(0)} = 0 \quad (1.4)$$

If the perturbed wavefunction is approximated as an expansion of the unperturbed wavefunctions (Equation 1.5),

$$\Psi_m^{(1)} = \sum_n c_n^{(1)}\Psi_n^{(0)} \quad (1.5)$$

then

$$\sum_n c_n^{(1)}(\mathcal{H}^{(0)} - E_m^{(0)})\Psi_m^{(0)} + (\mathcal{H}' - E_m^{(1)})\Psi_m^{(0)} = 0 \quad (1.6)$$

In Equation 1.5* the $\Psi_n^{(0)}$ are orthonormal, that is,

$$\int \Psi_n^{(0)}\Psi_{n'}^{(0)} \, d\tau = \delta_{nn'} \quad (1.7)$$

and

$$\int \Psi_n^{(0)}\mathcal{H}^{(0)}\Psi_{n'}^{(0)} \, d\tau = \delta_{nn'}E_{nn'}^{(0)} \quad (1.8)$$

So premultiplying Equation 1.6 by $\Psi_m^{(0)}$ and integrating gives the first-order correction to the energy,

$$E_m^{(1)} = \int \Psi_m^{(0)}\mathcal{H}'\Psi_m^{(0)} \, d\tau \quad (1.9)$$

In applying the results to a real chemical problem we may choose the value of the dummy parameter λ to be equal to unity such that $\mathcal{H} = \mathcal{H}^{(0)} + \mathcal{H}'$ and $E' = E^{(0)} + E^{(1)} + \cdots$. The first-order correction to the

---

* We shall always assume that our Ψ, ψ, or ϕ may be written as real functions and so shall not formally include the complex conjugate in such integrals.

wavefunction is simply obtained by premultiplying Equation 1.6 by all other unperturbed wavefunctions $\Psi_n^{(0)}$ except $\Psi_m^{(0)}$ and integrating such that

$$\Psi_m^{(1)} = \sum_n{}' \frac{\int \Psi_m^{(0)} \mathcal{H}' \Psi_n^{(0)} \, d\tau}{\Delta E_{mn}} \cdot \Psi_n^{(0)} \tag{1.10}$$

where the prime indicates that $n = m$ is excluded from the summation and $\Delta E_{mn} = E_m^{(0)} - E_n^{(0)}$.

The second-order correction to the energy (Equation 1.11) is obtained via extraction of terms from Equation 1.3 containing λ^2:

$$(\mathcal{H}^{(0)} - E_m^{(0)})\Psi_m^{(2)} + (\mathcal{H}' - E_m^{(1)})\Psi_m^{(1)} - E_m^{(2)}\Psi_m^{(0)} = 0 \tag{1.11}$$

Expanding $\Psi_m^{(2)}$ as above,

$$\Psi_m^{(2)} = \sum_n c_n^{(2)}\Psi_n^{(0)} \tag{1.12}$$

and substituting into Equation 1.5 gives

$$\sum_n [c_n^{(2)}(\mathcal{H}^{(0)} - E_m^{(0)})\Psi_n^{(0)} + c_n^{(1)}(\mathcal{H}'$$
$$- E_m^{(1)})\Psi_n^{(0)}] - E_m^{(2)}\Psi_m^{(0)} = 0 \tag{1.13}$$

Premultiplying by $\Psi_m^{(0)}$ and integrating as before gives

$$\sum_n{}' c_n^{(1)} \int \Psi_m^{(0)} \mathcal{H}' \Psi_n^{(0)} \, d\tau - E_m^{(2)} = 0 \tag{1.14}$$

which on substitution for $c_m^{(0)}$ gives

$$E_m^{(2)} = \sum_n{}' \frac{(\int \Psi_m^{(0)} \mathcal{H}' \Psi_n^{(0)} \, d\tau)^2}{\Delta E_{mn}} \tag{1.15}$$

Since the numerator is always positive, perturbation by states higher in energy than m leads to a negative contribution to the second-order energy, and perturbation by states lower in energy than m leads to a positive contribution. We use this result extensively in the discussion in this book. Higher-order perturbation results are obtained in a similar fashion. Sometimes we use these results as they stand, namely, to view the mixing of

electronic states Ψ as a result of a perturbation. More often we are interested in the mixing of atomic (ϕ) or molecular (ψ) orbitals as the result
of a perturbation, the results of which are approachable by exactly analogous mathematics by simply replacing Ψ by ψ or ϕ.

An alternative way exists of obtaining these results which has the
advantage of revealing what is neglected in our analysis above and also
of removing the assumption that the perturbation is small. Manipulation
of the wave equation (see Ref. 144 for details) leads to a secular determinant (Equation 1.16),

$$\begin{vmatrix} H_{11}' - (E - E_1^{(0)}) & H_{12}'... \\ H_{21}' & H_{22}' - (E - E_2^{(0)})... \\ \vdots & \vdots \end{vmatrix} = 0 \qquad (1.16)$$

which describes how the states or orbitals interact with each other when
a perturbation $\mathcal{H}'$ is applied to the system; H_{ij}' is the integral $\int \psi_i \mathcal{H}' \psi_j$
$d\tau$, and the roots of the determinant give the energy shifts $E_i^{(0)} - E_i$ of
each level i that result. As a first approximation to E_m (for example) we
can neglect all the elements in this determinant except H_{mm}', in which
case $E_m = E_m^{(0)} + H_{mm}'$, a result identical to the first-order correction
of Equation 1.9. By neglecting all elements of the determinant that do not
lie in the mth row or mth column, we arrive at the second-order result
of Equation 1.15. Inclusion of more elements of the secular determinant
leads to the results of higher-order perturbation theory. In the following
chapters we use both the secular determinant approach to calculate new
energy levels as the result of a perturbation and also the results of second-
order perturbation theory.

1.2   Overlap Integrals between Atomic Orbitals

To begin we consider some of the simple results of the solution of the
Schrödinger Wave Equation for atoms. In particular we focus on the form
of the wavefunctions and, as we are interested in chemical bonding, the
ways of expressing the overlap integral between two orbitals on different
atoms located in a specific geometric orientation with respect to each
other.

Solution of the Schrödinger wave equation for a single electron, charge
$-e$, and reduced mass μ moving in the potential produced by a nucleus

of charge $+Ze$ leads to a series of eigenvalues and eigenvectors (Equation 1.17) defined by the three quantum numbers n, l, and m:

$$E_n = -\frac{\mu Z^2 e^4}{2\hbar^2 n^2}$$

$$\psi_{nlm} = R_{nl}(r)\, Y_{lm}(\theta, \phi) \tag{1.17}$$

$$= R_{nl}(r)\Theta_{lm}(\theta)\, \Phi_m(\phi)$$

The variables r, θ, and ϕ are the conventional polar coordinates. The Y_{lm} are spherical harmonics, $R_{nl}(r)$ contains the associated Laguerre functions, and the exact analytic form of Equation 1.17 is readily written down. Note that the energy of the orbital, defined by n, l, and m quantum numbers, is independent of l and m and is determined by the value (in nonrelativistic theory at least) of n only. Thus $3s$ ($nl = 30$), $3p(nl = 31)$, and $3d(nl = 32)$ are equienergetic. $\Phi(\phi)$ is given by Equation 1.18:

$$\Phi_m = N\, e^{im\phi} \tag{1.18}$$

$N = 1/\sqrt{2\pi}$ is a normalization factor. To avoid the use of complex numbers, new functions are defined whenever $m \neq 0$. Thus

$$p_z\, \Phi_z = \Phi_0 = N \qquad (m = 0)$$

$$p_x\, \Phi_x = \frac{N}{\sqrt{2}} [\Phi_1 + \Phi_{-1}] = N \cos\phi$$

$$p_y\, \Phi_y = \frac{Ni}{\sqrt{2}} [\Phi_1 - \Phi_{-1}] = N \sin\phi$$

$$(m = \pm 1) \tag{1.19}$$

and their complete angular dependence is $\cos\theta$, $\sin\theta \cos\phi$, and $\sin\theta \sin\phi$, respectively.

For polyelectron atoms the Schrödinger wave equation becomes a many-body problem and cannot be solved exactly. Approximate solutions may be obtained by means which are referred to later. The most important result is that the wavefunctions take a similar form as before (Equation 1.20),

$$\psi_{nlm} = R'_{nl}(r)\, Y_{lm}(\theta, \phi) \tag{1.20}$$

but with the vital difference that $R'_{nl}(r)$ is not a simple analytic function and cannot be obtained exactly. The $Y_{lm}(\theta, \phi)$ are the same spherical harmonics as before and are of course readily determined. However for single-electron and polyelectronic atoms the overlap integral between two orbitals on different atoms may be written in a simple fashion, Equation 1.21, as a product of radial and angular terms:

$$S_{ab} = S_\lambda(r) \, F(\theta, \phi, \lambda) \tag{1.21}$$

$S_\lambda(r)$ depends upon the distance between the two atomic centers, the nature of the atoms a and b (i.e., whether Cd, S, or Na, for example), and the n and l values of both orbitals. It is also dependent upon whether the overlap has local $\lambda = \sigma, \pi,$ or δ symmetry. The angular term $F(\theta, \phi, \lambda)$ is a simple function of the angular polar coordinates of one atom relative to another, different functions being found for different values of λ. For example in Figure 1.1a the overlap integral between a p_z orbital on one atom and a σ type orbital on another is simply given by

$$S = S_\sigma \cos \theta \tag{1.22}$$

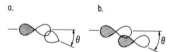

Figure 1.1 Overlap integrals between orbitals on two different atoms.

where S_σ contains all the radially dependent parts of the overlap integral and $\cos \theta$ all the angular dependence. Often for ease of calculation we need the overlap integral between two orbitals which are defined by a set of Cartesian axes (Figure 1.1b). In general the result is given by a sum of terms (Equation 1.23):

$$S = \sum_{\lambda=\sigma,\pi,\delta} S_\lambda F(\theta, \phi, \lambda) \tag{1.23}$$

In this particular example the overlap integral is simple $S = \sin^2 \theta \, S_\pi - \cos^2 \theta \, S_\sigma$. As will be seen on many occasions in this book, knowledge of the *angular* dependence only of the overlap integrals, which is the same irrespective of the chemical nature of *a* and *b* and the *ab* distance, is invaluable when looking at molecular structure. Tables 1.1 and 1.2 give values of the angular part of the overlap integrals between pairs of orbitals after the styles of Equations 1.21 (Figure 1.1a) and 1.23 (Figure 1.1b), respectively. Clearly the values in Table 1.1 are derived from those in Table 1.2 by a simple geometric transformation.

Table 1.1 Some Useful Overlap Integrals Between Central-Atom s, p. and d Orbitals and Ligand σ and π Orbitals[a,b]

$$S(s,\sigma) = S_\sigma$$
$$S(s,\pi) = 0$$
$$S(z,\sigma) = HS_\sigma$$
$$S(z,\pi_{\parallel}) = IS_\pi$$
$$S(z,\pi_{\perp}) = 0$$
$$S(z^2,\sigma) = \tfrac{1}{2}(3H^2 - 1)S_\sigma$$
$$S(x^2 - y^2,\sigma) = \sqrt{3}/2(F^2 - G^2)S_\sigma$$
$$S(xy,\sigma) = \sqrt{3}FGS_\sigma$$
$$S(xz,\sigma) = \sqrt{3}FHS_\sigma$$
$$S(yz,\sigma) = \sqrt{3}GHS_\sigma$$
$$S(z^2,\pi_{\parallel})^c = \sqrt{3}HIS_\pi$$
$$S(z^2,\pi_{\perp}) = 0$$
$$S(x^2 - y^2,\pi_{\parallel}) = -HIS_\pi$$
$$S(x^2 - y^2,\pi_{\perp}) = 0$$
$$S(xy,\pi_{\parallel}) = 0$$
$$S(xy,\pi_{\perp}) = IS_\pi$$
$$S(xz,\pi_{\parallel}) = (I^2 - H^2)S_\pi$$
$$S(xz,\pi_{\perp}) = 0$$
$$S(yz,\pi_{\parallel}) = 0$$
$$S(yz,\pi_{\perp}) = HS_\pi$$
$$F = \sin\theta\cos\phi$$
$$G = \sin\theta\sin\phi$$
$$H = \cos\theta$$
$$I = \sin\theta$$

[a] $\pi_{\parallel}$ is a ligand π orbital whose axis lies in a plane containing the z-axis and the ligand; $\pi_{\perp}$ is a ligand π orbital with an axis perpendicular to this plane.
[b] For p_z, d_{z^2}, f_{xyz}, etc. we use z, z^2, xyz, etc.
[c] Ligand lies in xz plane. For more general cases, manipulation of Table 1.2 is needed.

7

Table 1.2 Angular Dependence of Overlap Integrals[a,b]

$S(s,s) = S_\sigma$

$S(s,z) = -HS_\sigma$

$S(s,x) = -FS_\sigma$

$S(s,y) = -GS_\sigma$

$S(s,z^2) = (3H^2 - 1)S_\sigma/2$

$S(s,xz) = \sqrt{3}FHS_\sigma$

$S(s,yz) = \sqrt{3}GHS_\sigma$

$S(s,xy) = \sqrt{3}FGS_\sigma$

$S(s,x^2-y^2) = \sqrt{3}(F^2 - G^2)S_\sigma/2$

$S(s,z^3) = -H(5H^2 - 3)S_\sigma/2$

$S(s,xz^2) = -\sqrt{6}F(5H^2 - 1)S_\sigma/4$

$S(s,yz^2) = -\sqrt{6}G(5H^2 - 1)S_\sigma/4$

$S(s,xyz) = -\sqrt{15}FGHS_\sigma$

$S(s,z(x^2-y^2)) = -\sqrt{15}H(F^2 - G^2)S_\sigma/2$

$S(s,x(x^2-3y^2)) = -\sqrt{10}F(F^2 - 3G^2)S_\sigma/4$

$S(s,y(3x^2-y^2)) = -\sqrt{10}G(3F^2 - G^2)S_\sigma/4$

$S(z,z) = (1 - H^2)S_\pi - H^2S_\sigma$

$S(z,x) = -FH(S_\sigma + S_\pi)$

$S(z,y) = -GH(S_\sigma + S_\pi)$

$S(z,z^2) = -\sqrt{3}H(1 - H^2)S_\pi + H(3H^2 - 1)S_\sigma/2$

$S(z,xz) = -F(1 - 2H^2)S_\pi + \sqrt{3}FH^2S_\sigma$

8

$$S(z,yz) = -G(1 - 2H^2)S_\pi + \sqrt{3}GH^2 S_\sigma$$

$$S(z,xy) = FGH(\sqrt{3}S_\sigma + 2S_\pi)$$

$$S(z,x^2-y^2) = -H(G^2 - F^2)S_\pi + \sqrt{3}H(F^2 - G^2)S_\sigma/2$$

$$S(z,z^3) = \sqrt{6}(1 - H^2)(5H^2 - 1)S_\pi/4 - H^2(5H^2 - 3)S_\sigma/2$$

$$S(z,xz^2) = FH(11 - 15H^2)S_\pi/4 - \sqrt{6}FH(5H^2 - 1)S_\sigma/4$$

$$S(z,yz^2) = GH(11 - 15H^2)S_\pi/4 - \sqrt{6}GH(5H^2 - 1)S_\sigma/4$$

$$S(z,xyz) = \sqrt{10}FG(1 - 3H^2)S_\sigma/2 - \sqrt{15}FGH^2 S_\sigma$$

$$S(z,z(x^2-y^2)) = \sqrt{10}(F^2 - G^2)(1 - 3H^2)S_\pi/4 - \sqrt{15}H^2(F^2 - G^2)S_\sigma/2$$

$$S(z,x(x^2-3y^2)) = -FH(F^2 - 3G^2)(\sqrt{10}S_\pi + \sqrt{15}S_\pi)/4$$

$$S(z,y(3x^2-y^2)) = -GH(3F^2 - G^2)(\sqrt{10}S_\sigma + \sqrt{15}\,S_\pi)/4$$

$$S(x,x) = (1 - F^2)S_\pi - F^2 S_\sigma$$

$$S(x,y) = -FG(S_\sigma + S_\pi)$$

$$S(x,z^2) = \sqrt{3}FH^2 S_\pi + F(3H^2 - 1)S_\sigma/2$$

$$S(x,xz) = -H(1 - 2F^2)S_\pi + \sqrt{3}F^2 HS_\sigma$$

$$S(x,yz) = FGH(\sqrt{3}S_\sigma + 2S_\pi)$$

$$S(x,xy) = -G(1 - 2F^2)S_\pi + \sqrt{3}F^2 GS_\sigma$$

$$S(x,x^2-y^2) = -F(1 - F^2 + G^2)S_\pi + \sqrt{3}F(F^2 - G^2)S_\sigma/2$$

$$S(x,z^3) = -\sqrt{6}FH(5H^2 - 1)S_\pi/4 - FH(5H^2 - 3)S_\sigma/2$$

$$S(x,xz^2) = -(1 - F^2 - 5H^2 + 15F^2H^2)S_\pi/4 - \sqrt{6}F^2(5H^2 - 1)S_\sigma/4$$

$$S(x,yz^2) = FG(1 - 15H^2)S_\pi/4 - \sqrt{6}FG(5H^2 - 1)S_\sigma/4$$

$$S(x,xyz) = \sqrt{10}GH(1 - 3F^2)S_\pi/2 - \sqrt{15}F^2 GHS_\sigma$$

$$S(x,z(x^2-y^2)) = \sqrt{10}FH(2 - 3(F^2 - G^2))S_\pi/4 - \sqrt{15}FH(F^2 - G^2)S_\sigma/2$$

Table 1.2 (Continued)

$$S(x,x(x^2-3y^2)) = \sqrt{15}(F^2 - G^2 F^2(F^2 - 3G^2))S_\pi/4 - \sqrt{10}F^2(F^2 - 3G^2)S_\sigma/4$$

$$S(x,y(3x^2-y^2)) = \sqrt{15}FG(2 + G^2 - 3F^2)S_\pi/4 - \sqrt{10}FG(3F^2 - G^2)S_\sigma/4$$

$$S(y,y) = (1 - G^2)S_\pi - G^2 S_\sigma$$

$$S(y,z^2) = \sqrt{3}GH^2 S_\pi + G(3H^2 - 1)S_\sigma/2$$

$$S(y,xz) = FGH(\sqrt{3}S_\sigma + 2S_\pi)$$

$$S(y,yz) = -H(1 - 2G^2)S_\pi + \sqrt{3}G^2 HS_\sigma$$

$$S(y,xy) = -F(1 - 2G^2)S_\pi + \sqrt{3}FG^2 S_\sigma$$

$$S(y,x^2-y^2) = -G(G^2 - F^2 - 1)S_\pi + \sqrt{3}G(F^2 - G^2)S_\sigma/2$$

$$S(y,z^3) = \sqrt{6}HG(1 - 5H^2)S_\pi/4 - GH(5H^2 - 3)S_\sigma/2$$

$$S(y,xz^2) = FG(1 - 15H^2)S_\pi/4 - \sqrt{6}FG(5H^2 - 1)S_\sigma/4$$

$$S(y,yz^2) = -(1 - G^2 - 5H^2 + 15G^2 H^2)S_\pi/4 - \sqrt{6}G^2(5H^2 - 1)S_\sigma/4$$

$$S(y,xyz) = \sqrt{10}FH(1 - 3G^2)S_\pi/2 - \sqrt{15}FG^2 HS_\sigma$$

$$S(y,z(x^2-y^2)) = -\sqrt{10}GH(2 + 3(F^2 - G^2))S_\pi/4 - \sqrt{15}GH(F^2 - G^2)S_\sigma/2$$

$$S(y,x(x-3y^2)) = -\sqrt{15}FG(2 + F^2 - 3G^2)S_\pi/4 - \sqrt{10}FG(F^2 - 3G^2)S_\sigma/4$$

$$S(y,y(3x^2-y^2)) = \sqrt{15}(F^2 - G^2 - G^2(3F^2 - G^2))S_\pi/4 - \sqrt{10}G^2(3F^2 - G^2)S_\sigma/4$$

$$S(z^2,z^2) = 3(1 - H^2)^2 S_\pi/4 - 3H^2(1 - H^2)S_\pi + (3H^2 - 1)^2 S_\sigma/4$$

$$S(z^2,xz) = \sqrt{3}FH(H^2 - 1)S_\pi/2 - \sqrt{3}FH(1 - 2H^2)S_\pi - \sqrt{3}FH(3H^2 - 1)S_\sigma/2$$

$$S(z^2,yz) = \sqrt{3}GH(H^2 - 1)S_\pi/2 - \sqrt{3}GH(1 - 2H^2)S_\pi + \sqrt{3}GH(3H^2 - 1)S_\sigma/2$$

$$S(z^2,xy) = \sqrt{3}FG(1 + H^2)S_\pi/2 + 2\sqrt{3}FGH^2 S_\pi + \sqrt{3}FG(3H^2 - 1)S_\sigma/2$$

$$S(z^2, x^2-y^2) = (G^2 + F^2H^2)(1 + H^2)S_\delta/4 - \sqrt{3}H^2(G^2 - F^2)S_\pi + \sqrt{3}(F^2 - G^2)(3H^2 - 1)S_\sigma/4$$

$$S(xz, xz) = (G^2 + F^2H^2)S_\delta - (1 - 4F^2H^2 - G^2)S_\pi + 3F^2H^2S_\sigma$$

$$S(xz, yz) = FG(H^2 - 1)S_\delta - FG(1 - 4H^2)S_\pi + 3FGH^2S_\sigma$$

$$S(xz, xy) = GH(F^2 - 1)S_\delta - GH(1 - 4F^2)S_\pi + 3F^2GHS_\sigma$$

$$S(xz, x^2-y^2) = FH(F^2 - G^2 - 2)S_\delta/2 - FH(1 - 2F^2 + 2G^2)S_\pi + 3FH(F^2 - G^2)S_\sigma/2$$

$$S(yz, yz) = (F^2 + G^2H^2)S_\delta - (1 - F^2 - 4G^2H^2)S_\pi + 3G^2H^2S_\sigma$$

$$S(yz, xy) = FH(G^2 - 1)S_\delta - FH(1 - 4G^2)S_\pi + 3FG^2HS_\sigma$$

$$S(yz, x^2-y^2) = GH(F^2 - G^2 + 2)S_\sigma/2 - GH(2G^2 - 2F^2 - 1)S_\pi + 3GH(F^2 - G^2)S_\sigma/2$$

$$S(xy, xy) = (H^2 + F^2G^2)S_\delta - (1 - H^2 - 4F^2G^2)S_\pi + 3F^2G^2S_\sigma$$

$$S(xy, x^2-y^2) = FG(F^2 - G^2)S_\delta/2 - 2FG(G^2 - F^2)S_\pi + 3FG(F^2 - G^2)S_\sigma/2$$

$$S(x^2-y^2, x^2-y^2) = (4H^2 + (F^2 - G^2))S_\delta/4 - (1 - H^2 - (F^2 - G^2)^2)S_\pi + 3(F^2 - G^2)^2S_\sigma/4$$

Adapted from Ref. 196.

[a] For p_z, d_z^2, f_{xyz}, and so on, we use z, z^2, xyz, and so on.

[b] The atom holding the first orbital is situated at the origin, and the atom containing the second orbital is defined by the direction cosines F, G, H. The local x-, y-, z-axes on each center are codirectional. This means, for example; that the + lobe of a z orbital points in the $+z$ direction for both orbitals of the table.

$F = \sin\theta\cos\phi$

$G = \sin\theta\sin\phi$

$H = \cos\theta$

1.3   Properties of Molecular Orbitals

If two atomic orbitals, ϕ_i and ϕ_j, on two different atoms r and s (or molecular orbitals on two molecular fragments) with energies E_i and E_j are allowed to interact, then the second-order perturbation theory result of Section 1.1 tells us that the lower-energy orbital ϕ_i will be depressed lower in energy and the higher-energy one pushed to higher energy (Figure 1.2). Consideration of Equation 1.10 shows that the higher-energy orbital is still largely ϕ_i and the deeper-lying orbital is still largely ϕ_j. From the sign of the energy denominator in Equation 1.10 we can see that ϕ_i and ϕ_j are mixed in phase in the lower-energy (bonding) orbital but out of phase in the higher-energy antibonding orbital. In polyatomic molecules in general, where molecular orbitals are usually expressed as a linear combination of atomic orbitals (LCAO), as in Equation 1.24,

$$\psi_m = \sum_i c_{im}\phi_i \tag{1.24}$$

the orbital energy increases (i.e., becomes less stable) as the number of nodes in ψ_m increases. In this simple two-orbital case the deepest-lying orbital (in-phase bonding) has no nodes and the higher-lying orbital (out-of-phase, antibonding) has one node.

Mulliken has identified electronegativity with the mean of electron affinity and ionization potential. Thus, in simple terms, the more negative a valence orbital E_j the more electronegative the atomic species. We can see how this arises in the following way. If the bonding orbital (ψ_b) between ϕ_i and ϕ_j is written

$$\psi_b = c_{ib}\phi_i + c_{jb}\phi_j \tag{1.25}$$

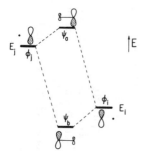

Figure 1.2   The character of molecular orbitals arising from overlap of atomic orbitals.

then, since $|E_i| > |E_j|$ and $c_{ib} > c_{jb}$, the charge distribution

$$\psi_b^2 = c_{ib}^2 \phi_i^2 + c_{jb}^2 \phi_j^2 + 2c_{ib}c_{jb}\phi_i\phi_j \tag{1.26}$$

shows that the electron density is largely located in the orbital ϕ_i ($c_{ib}^2 > c_{jb}^2$). This is the atomic orbital with the larger E and hence corresponds to the more electronegative atom. The reverse result occurs for the antibonding orbital ψ_a:

$$\psi_a = c_{ia}\phi_i + c_{ja}\phi_j \tag{1.27}$$

Here $c_{ia} < c_{ja}$ and the electron density is largely located (Figure 1.2) on the least electronegative atom. Figure 1.3 shows the experimental ^{216}Ap orbital density in nonlinear 19-electron radicals AX$_2$(A = N, P; X = F, Cl) obtained via electron paramagnetic resonance (epr). The unpaired electron occupies an orbital that is AX antibonding, and clearly the electron density in the Ap orbital located on the least electronegative atom increases as the AX electronegativity difference increases, just as expected.

We may use this simple two-orbital model to illustrate a general way of obtaining a measure of the bond "strength" or bond order in molecules and the atomic charges from molecular orbital calculations. If the kth molecular orbital is written

$$\psi_k = c_{ik}\phi_i + c_{jk}\phi_j \tag{1.28}$$

and contains N_k electrons, the charge distribution is simply

$$N_k\psi_k^2 = N_k c_{ik}^2 \phi_i^2 + N_k c_{jk}^2 \phi_j^2 + 2N_k c_{ik}c_{jk}\phi_i\phi_j \tag{1.29}$$

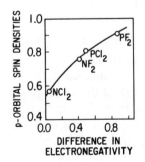
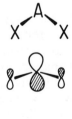

Figure 1.3 A-atom spin density in 19-electron AX$_2$ radicals as a function of AX electronegativity difference (from Ref. 216).

On integration this yields

$$N_k = N_k c_{ik}^2 + N_k c_{jk}^2 + 2N_k c_{ik} c_{jk} S_{ij} \qquad (1.30)$$

where S_{ij} is the overlap integral between the orbitals i and j, and $N_k c_{ik}^2$ is the contribution to the net atomic population in orbital i by the electrons in the kth molecular orbital. $P_{rs}(k) = 2N_k c_{ik} c_{jk} S_{ij}$ is the contribution via occupation of the kth orbital to the overlap population between the two atoms r and s on which the orbitals i and j are located. The total bond overlap population between the two atoms r and s in the molecule is the sum (Equation 1.31)

$$P_{rs} = 2 \sum_{i,j} \sum_{k} N_k c_{ik} c_{jk} S_{ij} \qquad (1.31)$$

of this function over all the molecular orbitals (clearly only occupied orbitals will contribute to the summation) and over all possible pairs of orbitals i and j located on the two atoms. The bond overlap population P_{rs}, is often used as a measure of bond strength.

The contribution to the gross atomic population by occupation of orbital i, $q_r(i)$, is obtained by adding to the net atomic population a contribution from the overlap population such that

$$q_r(i) = \sum_{j} \sum_{k} (N_k c_{ik}^2 + N_k c_{ik} c_{jk} S_{ij}) \qquad (1.32)$$

where the sum is over all molecular orbitals k and all atomic orbitals j for which the overlap integral S_{ij} is nonzero. When summed over all orbitals on the atom r, the total gross atomic population is found:

$$q_r = \sum_{i} q_r(i) \qquad (1.33)$$

The net atomic charge on atom r is then the difference between q_r and the number of electrons in a neutral free r atom.

The result of perturbation theory (Equation 1.10) implies that atomic orbitals will only mix together strongly if their energy separation is small. Thus we are usually only interested in the interactions of valence orbitals on one center with valence orbitals on another rather than with valence orbital overlap with the deeper-lying core orbitals or higher-energy unoccupied orbitals of the next shell. Alternatively two atomic orbitals will only make significant contributions to the same molecular orbital if their energies before interaction are comparable. The normalization procedure itself also determines the relative sizes of the coefficients that we illustrate with the two-orbital system. Normalization of Equation 1.25 gives

$$c_{ib}^2 + c_{jb}^2 + 2c_{ib}c_{jb}S_{ij} = 1 \qquad (1.34)$$

Since this is the bonding component, the product $c_{ib}c_{jb}S_{ij} > 0$, and so $c_{ib}^2 + c_{jb}^2 < 1$. This implies that both c_{ib} and c_{jb} are smaller than unity. For the antibonding orbital (Equation 1.27), $c_{ia}c_{ja}S_{ij} < 0$, and so $c_{ia}^2 + c_{ja}^2 > 1$. The possibility now arises that c_{ia} or c_{ja} or both may be larger than unity. This is a general trend. As a stack of molecular orbitals are climbed, an increasing number of greater-than-unity coefficients are encountered.

One final result[93] we may extract from the perturbation theory approach of Section 1.1 is that via Equation 1.15 the energy shift associated with orbital m when two fragments or atoms are brought together is pairwise additive. In other words, the total energy shift is the sum of that arising from the interaction between orbital m and orbital 1, plus that arising from the interaction between orbital m and orbital 2, and so on.

2 Approximate Molecular Orbital Methods

Since most of the models discussed in this book are based on simple molecular orbital ideas, it is necessary to appreciate the origin of these approximate orbital methods. It will quickly be realized that the approaches used are ruthless simplifications of quantitative, "exact" molecular orbital theory.

2.1 Self-Consistent Field Theory[48,58,174,191]

The form of the wavefunction describing the behavior of spin one-half particles (e.g., electrons) is governed by the operation of the antisymmetry (Pauli) principle. The total electronic wavefunction Ψ is required to be antisymmetric with respect to the exchange of any pair of electrons. Thus for two electrons labeled i and j located in spin orbitals ψ_a and ψ_b, Ψ is given by Equation 2.1, where $\mathcal{A}$ is an antisymmetrising operator:

$$\Psi = \mathcal{A}\psi_a(i)\,\psi_b(j)$$

$$= \frac{1}{\sqrt{2}}\,[\psi_a(i)\,\psi_b(j) - \psi_a(j)\,\psi_b(i)] \tag{2.1}$$

Equation 2.1 changes sign if the labels i and j are exchanged. In the many-electron case, Ψ is best written as a Slater determinant (Equation 2.2). For the case of n electrons,

$$\Psi = \frac{1}{\sqrt{n!}} \begin{vmatrix} \psi_a(1)\,\psi_a(2).....\psi_a(n) \\ \psi_b(1)\,\psi_b(2).....\psi_b(n) \\ \vdots \quad \vdots \quad \vdots \\ \psi_n(1)\,\psi_n(2).....\psi_n(n) \end{vmatrix} \tag{2.2}$$

The molecular orbitals, ψ_i, describe the behavior of an electron in a molecule subject to a potential due to the nuclei and all the other electrons. They are eigenfunctions of a one-electron Hamiltonian. The interaction between the jth electron and all the others may be described by an average potential such that

$$\mathcal{H}_F \psi_i = E_i \psi_i \qquad (2.3)$$

where $\mathcal{H}_F$ is the Fock operator which for closed-shell systems may be written as

$$\mathcal{H}_F = \mathcal{H}_C + \sum_{k=1}^{n} (2\mathcal{J}_k - \mathcal{H}_k) \qquad (2.4)$$

The "orbital energies" are the eigenvalues (E_i) of this operator. $\mathcal{H}_c$ is the "core" Hamiltonian and contains the one-electron terms, the electron–nuclear attraction, and the kinetic energy operator. The other term contains the two-electron functions $\mathcal{J}$ and $\mathcal{H}$, operators defined by Equations 2.5:

$$\mathcal{J}_1 \psi_i(1) = \psi_i(1) \int \psi_j(2) \frac{1}{r_{12}} \psi_j(2)\, d\tau_2$$
$$\mathcal{H}_1 \psi_i(1) = \psi_j(1) \int \psi_j(2) \frac{1}{r_{12}} \psi_i(2)\, d\tau_2 \qquad (2.5)$$

On integration, these lead to J_{ij}, the Coulomb repulsion integral, and K_{ij}, the exchange integral (Equations 2.6):

$$J_{ij} = \int \psi_a(i)\, \psi_b(j) \frac{e^2}{r_{ij}} \psi_a(i)\, \psi_b(j)\, d\tau$$
$$K_{ij} = \int \psi_a(i)\, \psi_b(j) \frac{e^2}{r_{ij}} \psi_a(j)\, \psi_b(i)\, d\tau \qquad (2.6)$$

J_{ij} represents the classical coulombic repulsion of an electron i in orbital ψ_a with electron j in orbital ψ_b, and its origin is readily appreciated. The exchange integral K_{ij} derives its name simply because the i, j labels on the right-hand side of the operator e^2/r_{ij} are exchanged compared to the Coulomb integral. Its origin lies exclusively in the domain of quantum mechanics and arises via the antisymmetry principle. Recall that Equation 2.1 contained two terms, the second with the i, j labels reversed compared to the first. An important physical consequence of its presence is that systems containing paired electron spins lie higher in energy than systems with the same orbital occupancy but unpaired spins. The exchange integral can be envisaged as the energy term that prevents electrons of the same spin from occupying the same region of space.

Clearly the $\mathcal{J}$ and $\mathcal{K}$ of Equation 2.4 involve the form of the orbitals ψ that need to be determined in Equation 2.3. This equation then needs to be solved in an iterative fashion. For a zeroth order set of orbitals ψ (i.e., a guess or the result of an approximate calculation), the operators are defined from which a new set of orbitals are determined by solution of Equation 2.3. These orbitals are used to produce new values for the operators, and Equation 2.3 is solved again. The process is repeated until convergence, that is, when the Hamiltonian produced at the end of a cycle is consistent with the orbitals that produced it in the first place. $\mathcal{H}_F$ is then a self-consistent field (SCF) Hamiltonian. In molecular orbital calculations a commonly used approximation is to expand the ψ_i of Equation 2.3 as a linear combination of atomic orbitals ϕ_i (Equation 2.7):

$$\psi_j = \sum_i^n c_{ij}\phi_i \tag{2.7}$$

If a large enough number of basis functions (atomic orbitals) are used, then exact SCF molecular orbitals may be obtained by adjusting the c_{ij} in each cycle, and the energies would be close to the so-called Hartree–Fock limit. These are the energies that would be obtained via the SCF procedure if there were no restriction on the sorts of function the ψ_j could adopt. Usually a rather restricted set of atomic orbitals is included in Equation 2.7, and best orbital energies are obtained by a method due to Roothaan by applying the variation principle. The Roothaan–Hartree–Fock method leads to values for the c_{ij} in Equation 2.7, which may be used to determine the energies from Equation 2.3. The ϕ_i used in Equation 2.7 are the best atomic wavefunctions obtained by applying very similar SCF procedures to atoms rather than molecules. As we saw in Section 1.2, it is the radial part of the wavefunction that is responsible for energy variations of a given orbital from one atom to another. Thus by expansion of $R(r)$ in terms of a readily available basis set, Slater orbitals[195] (see Section 2.4) as in Equation 2.8,

$$R_{nl}(r) = \sum_\zeta c_\zeta r^{n-1} e^{-\zeta r} \tag{2.8}$$

or Gaussians (where the radial dependence is $r^{2n} e^{-\zeta r^2}$), the c_ζ may be obtained for a series of values of ζ. The "best" atomic wavefunctions include four or five Slater-type orbitals (STOs) or other basis per orbital; some others are of double-zeta (two functions) quality. (After all this work however, even "modified" versions of these orbital wavefunctions may actually be used[149] in the molecular calculation.) The end result is the eigenvalues and eigenfunctions of $\mathcal{H}_F$ of Equation 2.3, which are respectively the orbital energies and molecular orbital wavefunctions of the system.

All electron SCF calculations are then in general rather complex undertakings. In order to obtain energies close to the Hartree–Fock limit, a sizable number of functions need to be included in the expansions of Equations 2.7 and 2.8. This means that the number of integrations involved, both in the determination of the two-electron operators of Equation 2.4 and the eigenvalues of Equation 2.5, may be sizable. As an example, a calculation of this type for the 86-electron NiF_6^{-4} species might need around 10^8 integrations. It should be no surprise to find the extensive use of restricted basis sets when studying larger molecules. We shall not however discuss quality numerical calculations any further but move immediately to the other extreme, the approximate molecular orbital models. Although these represent a tremendous simplification over what has been outlined above, the results they can produce quickly have provided an understanding at a very basic level of a wide spectrum of chemical phenomena.

2.2 Approximate Methods

The approximate methods that are available at present[49,63,150,191,194,204] may be divided into two types—those that retain the SCF methodology but parametrize many of the various integrals needed (CNDO, MINDO, etc.[48,50,122]) and those which do not.[90,96,97,203,223] We discuss the latter type exclusively, not because the "NDO" methods (neglect of differential overlap) have little merit but because our arguments will be based on orbital overlap and, at the qualitative level, are best reached by modification of a noniterative method. Although the CNDO and related methods are fast and are successful in reproducing geometric and spectroscopic properties of organic molecules, they have not been used extensively in metal-containing systems. Other theoretical methods for the generation of good molecular orbitals and energies such as the more recently developed $X\alpha$ scattered wave method[39,107] do not immediately lend themselves to the sort of drastic simplification we need to employ to produce simple models whose workings are transparent to the average chemist at the semiquantitative level.

We saw in Section 2.1 that the Hamiltonian for a many-electron system could be replaced by the Fock operator (in Equation 2.9 we drop the subscript F):

$$\mathcal{H}\psi = E\psi \tag{2.9}$$

If we use the LCAO approximation, then

$$\mathcal{H}\left(\sum_i c_i\phi_i\right) = E\left(\sum_i c_i\phi_i\right) \tag{2.10}$$

and

$$E = \frac{\int\left(\sum_i c_i\phi_i\right)\mathcal{H}\left(\sum_i c_i\phi_i\right)d\tau}{\int\left(\sum_i c_i\phi_i\right)^2 d\tau} \tag{2.11}$$

Equation 2.11 may readily be rewritten as Equation 2.12,

$$E = \frac{\sum_j\sum_i c_ic_jH_{ik}}{\sum_j\sum_i c_ic_jS_{ik}} \tag{2.12}$$

where we introduce the following abbreviations (all ϕ_i assumed real functions):

$$H_{ii} = \int \phi_i\mathcal{H}\phi_i\,d\tau \qquad \text{Coulomb integral}$$

$$H_{ij} = \int \phi_i\mathcal{H}\phi_j\,d\tau \qquad \text{resonance integral} \tag{2.13}$$

$$S_{ij} = \int \phi_i\phi_j\,d\tau \qquad \text{overlap integral}$$

H_{ii}, the Coulomb integral, is not to be confused with J_{ij}, the Coulomb repulsion integral of Equation 2.6. By applying the variation theorem to Equation 2.12 we set each of the derivatives $\partial E/\partial c_i$ equal to zero. The result is a set of secular equations (Equations 2.14),

$$c_1(H_{11} - S_{11}E) + c_2(H_{12} - S_{12}E) + \cdots c_n(H_{1n} - S_{1n}E) = 0$$

$$c_1(H_{21} - S_{21}E) + c_2(H_{22} - S_{22}E) + \cdots c_n(H_{2n} - S_{2n}E) = 0 \tag{2.14}$$

$$\vdots \qquad\qquad\qquad \vdots \qquad\qquad\qquad \vdots$$

$$c_1(H_{n1} - S_{n1}E) + c_2(H_{n2} - S_{n2}E) + \cdots c_n(H_{nn} - S_{nn}E) = 0$$

which will be mathematically consistent only if the secular determinant (Equation 2.15) is equal to zero:

$$|H_{ij} - S_{ij}E| = 0 \qquad (2.15)$$

Clearly Equation 2.15 is similar to the secular determinant of Equation 1.16 with the modification that the basis set in this case is not orthogonal and the terms $S_{ij}E$ need to be included in the off-diagonal elements as a consequence. Solution of the secular equations gives a set of eigenvalues (E_j) and eigenvectors (c_{ij}). Thus if we can make an estimate of the H_{ii} (perhaps representing the energy of ϕ_i in an atomic environment) and H_{ij} (which may be viewed as the interaction energy between ϕ_i and ϕ_j), then, with computed values of S_{ij}, we will have a numerical estimate of a quantity that approximates to the energy of a molecular orbital energy level. In this way we shall have sidestepped the SCF procedure by collecting all the one- and two-electron energies into a single expression. We shall consider three simple approaches that derive from this viewpoint.

2.3   The Hückel Method[89,198,203,220]

This is the simplest approach of all and is used mainly to generate π levels in molecules, especially organic molecules containing extended π systems. Combined with perturbation theory, Hückel molecular orbital (HMO) theory is a powerful way to view organic chemistry.[50,151,184] In its simplest form in molecules containing carbon atoms alone, all $S_{ij} = 0$ for i and j on different atoms, the H_{ii} are parametrized as $H_{ii} = \alpha$, and all $H_{ij} = \beta$ when i and j are on adjacent atoms bonded via the σ framework; $H_{ij} = 0$ if the atoms are not so connected. For the ethylene molecule there are two π-type orbitals, ϕ_1 and ϕ_2, one on each carbon atom. The secular determinant becomes

$$\begin{vmatrix} \alpha - E & \beta \\ \beta & \alpha - E \end{vmatrix} = 0 \qquad (2.16)$$

which has roots $E = \alpha \pm \beta$. Here of course $H_{ij} = H_{ji}$ and $H_{ii} = H_{jj}$ by symmetry. Now both α and β are by definition negative, and so we may readily draw a simple molecular orbital diagram showing the result of the Hückel calculation where ψ_1 is the in-phase bonding combination to lower energy than ψ_2, the out-of-phase antibonding combination (Figure 2.1a).

If we do wish to include overlap in the scheme, one way of doing this

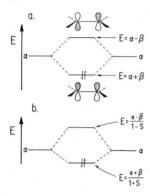

Figure 2.1 Hückel π orbitals of ethylene (*a*) without and (*b*) with overlap included between the two $p\pi$ orbitals.

is to properly normalize the (symmetry determined) ψ_i such that $\int \psi_i \psi_i \, d\tau = 1$. This gives the values in Equations 2.17:

$$\psi_1 = \frac{1}{\sqrt{2(1 + S)}} (\phi_1 + \phi_2)$$

$$\psi_2 = \frac{1}{\sqrt{2(1 - S)}} (\phi_1 - \phi_2)$$

(2.17)

Using $E_i = \int \psi_i \mathcal{H} \psi_i \, d\tau$ the energy levels of Figure 2.1*b* are obtained. Exactly the same result is obtained if Equation 2.15 is solved with overlap included. The important result to note is that the bonding orbital is stabilized less than the antibonding orbital is destabilized.

A slightly more complex example is the allyl anion $C_3H_5^-$ where the secular determinant (without overlap) is

$$\begin{vmatrix} \alpha - E & \beta & \beta \\ \beta & \alpha - E & 0 \\ \beta & 0 & \alpha - E \end{vmatrix} = 0$$

The roots are $E = \alpha, \alpha \pm \sqrt{2}\beta$. The coefficients shown in Figure 2.2 are obtained from these energies by manipulation of the secular equations. The electron density on a particular atom is given by the sum of the squares of the coefficients over all the occupied orbitals. The net and gross charges (Section 1.3) are equal in this illustration since we neglect the overlap integrals between the atomic orbitals. For the allyl radical with three π electrons, the electron density on the end atoms is the same as on the middle; but in the allyl negative ion with four π electrons, the end atoms carry a higher charge (one and a half electrons) compared to

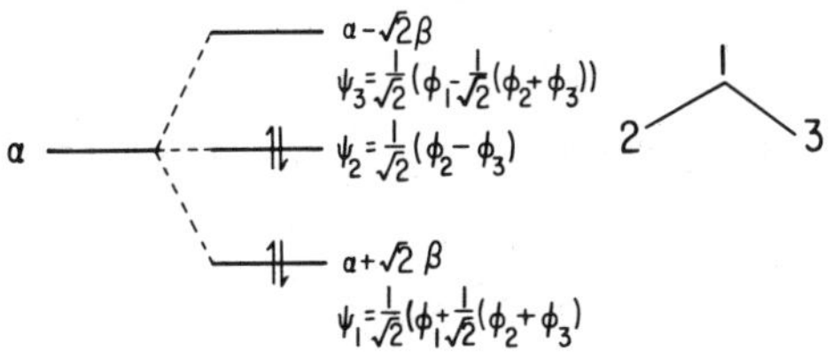

$$\alpha - \sqrt{2}\beta$$
$$\psi_3 = \tfrac{1}{\sqrt{2}}(\phi_1 - \tfrac{1}{\sqrt{2}}(\phi_2 + \phi_3))$$

$$\psi_2 = \tfrac{1}{\sqrt{2}}(\phi_2 - \phi_3)$$

$$\alpha + \sqrt{2}\,\beta$$
$$\psi_1 = \tfrac{1}{\sqrt{2}}(\phi_1 + \tfrac{1}{\sqrt{2}}(\phi_2 + \phi_3))$$

Figure 2.2   Hückel orbitals of the allyl anion.

the middle one (one electron). This is a general result for a $3c\text{–}4e$ (three-center–four-electron) bonding situation that we shall use later. A useful result concerning the Hückel energies of linear chains of n atoms is that the orbital energies may be written in the simple form $E_j = \alpha - 2\beta \cos (j\pi/(n + 1))$, where $j = 1, 2, 3, \ldots n$.

In organic systems the β parameter is approximately transferable from one molecule to another. The carbon atoms are roughly neutral in all cases and in similar electronic environments. Clearly the π stabilization energy in ethylene (with two electrons in ψ_1) is 2β, and in the allyl negative ion it is $2\sqrt{2}\beta$. In valence bond language the latter species may be described by the two resonating canonical forms $CH_2{=}CH{-}CH_2^-$ and $CH_2^-{-}CH{=}CH_2$, each containing a double bond. The molecular orbital delocalization energy, the difference in π stabilization energy between the allyl ion and one isolated double bond, $2(\sqrt{2} - 1)\beta$, may then be directly compared to the valence bond "resonance energy." Thermochemical measurements of the hydrogenation of these and other molecules lead to the determination of an empirical resonance energy in extended π systems. The correlation between these empirical values and the π delocalization energy calculated using the Hückel method is an impressive one (Figure 2.3). Not so impressive is the fact that different values of β are calculated depending upon the nature of the physical observable used.[150] Thus in Figure 2.3 the π-delocalization energy gives a value of $\beta = -0.69$ eV. Good predictive correlation is found for the absorption maxima of aromatic systems, but here $\beta = -2.71$ eV. Similarly polarographic half-wave reductions of aromatic hydrocarbons are well fitted but with $\beta = -2.37$ eV. So the physical observable used in such correlations determines the details of how the terms, ignored in these simple treatments, actually are absorbed into the parameters of the method.[63] The Hückel method may be used for atoms other than carbon, but here the transferability of the results is usually not quite so good. The method may be modified by the addition of more parameters (the ω technique)[220] to take into account the presence of heteroatoms.

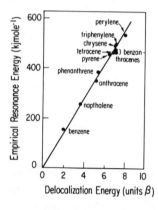

Figure 2.3. Plot of calculated "resonance energy" (units of β) against thermochemical data.

A striking application of HMO theory is in the $4n + 2$ rule of Hückel. Monocyclic coplanar rings of trigonaly coordinated atoms which contain $4n + 2$ π electrons are found to possess relative electronic stability. The π molecular orbitals of such rings may be written in a simple analytic form. For a ring containing n atoms the energies are given by $E_j = \alpha - 2\beta \cos(2j\pi/n)$, where $j = 0, \pm 1, \pm 2$, and so on. The results are shown graphically for cyclic systems of m atoms in Figure 2.4. An m-fold regular polygon is inscribed within a circle of radius 2β such that one apex is at the lowest point. The vertical distance of an apex from a horizontal line drawn through the center of the circle then gives an energy level in units of β. The lowest-lying MO is always single and can accomodate two electrons. Each of the remaining low-energy orbitals occur in pairs, and each pair is filled by four electrons. Thus closed shells of low-lying orbitals are found for $4n + 2$ electrons.

Since we use this method very little in the rest of the book, let us describe one of its applications in the structural field. The allene molecule $CH_2 \cdot C \cdot CH_2$ and its inorganic analog $(Me_3Si)_2NBeN(SiMe_3)_2$ have a D_{2d} ground-state geometry (Figure 2.5) where the planes of the CH_2 or NSi_2 groups attached to the central C or Be atom are perpendicular to one another. The planar D_{2h} geometry is not found. There are four $p\pi$-type

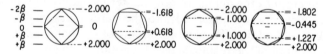

Figure 2.4 Hückel orbital energies of cyclic systems.

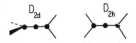

Figure 2.5 D_{2d} and D_{2h} geometries for $CH_2 \cdot C \cdot CH_2$ and $(Me_3Si)_2$ N·Be·N(SiMe_3)_2.

orbitals to be considered here, one on each of the CH_2 groups and two mutually perpendicular ones on the central atom. In the staggered D_{2d} geometry the situation is very much like that in the ethylene molecule. There are two ethylene-type π interactions perpendicular to each other (Figure 2.6). These are degenerate (species e). With four electrons the total stabilization energy is 4β. For the planar form there is one extended set of orbitals and one localized one. For the former, the secular determinant is identical to that for the allyl species (Equation 2.18) with roots $E = \alpha$, $\alpha \pm \sqrt{2}\beta$. The energy of the localized orbital is simply α since it is of the wrong symmetry to interact with any ligand orbital (Figure 2.7). The total stabilization energy of this isomer is $2\sqrt{2}\beta$ (= 2.828β), smaller than the 4β found for the D_{2d} geometry. The staggered geometry is thus the one predicted theoretically.

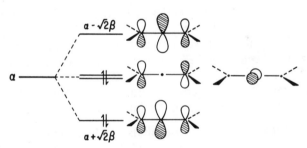

Figure 2.6 Hückel orbitals of D_{2h} $CH_2 \cdot C \cdot CH_2$.

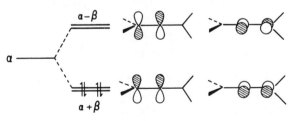

Figure 2.7 Hückel orbitals of D_{2d} $CH_2 \cdot C \cdot CH_2$.

2.4   The Extended Hückel Method[96,97,223]

The extended Hückel molecular orbital (EHMO) method has been widely used in both inorganic and organic chemistry, in the former sometimes under the name of the Wolfsberg–Helmholz method,[223] and is the only one of the three simple methods we shall discuss where numerical values are used in the evaluation of orbital energies. The method has formed the cornerstone of some exceedingly illuminating analyses of the factors influencing the geometries of transition metal systems by the Hoffmann school.[141] In the secular determinant (Equation 2.15) there are three quantities to be included, H_{ii}'s, H_{ij}'s and S_{ij}'s.

1. The S_{ij} are evaluated for a particular molecular geometry by using Slater-type orbitals for the ϕ_i. STO's contain a simplified radial wavefunction, $R(r)$, given[195] by Equation 2.19:

$$R(r) = r^{n^* - 1} e^{-\zeta r} \tag{2.19}$$

$$\zeta = \frac{Z - \Sigma S}{n^*}$$

where the values of n^* are related to n the principal quantum number in the following way. For $n = 1, 2, 3, 4, 5, 6$, $n^* = 1, 2, 3, 3.7, 4.0, 4.2$ (the difference is the quantum defect); Z is the nuclear charge; and S is the screening of the nucleus experienced by a valence electron by the other electrons in the same shell or by electrons lying closer to the nucleus. ζ is obtained by adding contributions to S from all the other electrons in the atom in the following way:

a.   All electrons in ns or np orbitals shield an electron in an ns or np orbital with $S = 0.35$ (a better value for an electron in a $1s$ orbital is $S = 0.30$).

b.   All electrons in the $n - 1$ shell shield with $S = 0.85$.

c.   All electrons in the $n - 2$ shell or deeper shield completely, that is, $S = 1.00$.

d.   For the shielding of a d or f electron all electrons closer to the nucleus shield completely ($S = 1$).

The STO's are certainly a poor representation of the radial wavefunction close to the nucleus and are probably not very good further away. That they are crude may be shown by the variation in values for the exponents for the $2s$, $2p$ orbitals in carbon optimized in various ways: 1.625, 1.625 (Slater,[195] from the rules above); 1.608, 1.568

(Clementi and Raimondi[38]); 1.550, 1.325 (Burns[33]); 1.60, 1.43 (Cusachs et al.[43]). Tabulations of Slater-type overlap integrals are found in Ref. 148.

2. The H_{ii} are put equal to the valence state ionization potential [6,14,136] (VSIP) or valence orbital ionization potential (VOIP). The VSIP or VOIP is a function of the charge and configuration of the atom. Spectral data[143] for the free gaseous atom or ion give data for the ionization potential (e.g., $2p$ in carbon) as a function of charge $q(C, C^+, C^{+2},$ etc.). Values for nonintegral charges may be obtained by fitting these values to an equation of the form of 2.20;

$$H_{ii} = Aq^2 + Bq + C \qquad (2.20)$$

and interpolating. (Some values of A, B, and C for a selection of species are given in Ref. 136.) The appropriate value is chosen for insertion into the secular determinant or an iterative procedure is adopted such that the value of H_{ii} is set by the computed atomic charge from the molecular orbital calculation. The calculation is repeated with a new value of H_{ii} from Equation 2.20 until self consistency is reached. This iterative method is called the Self-Consistent Charge and Configuration Method (SCCC). In most calculations however a fixed value of the VSIP is included and the secular determinant diagonalized just once. Some typical values of H_{ii} are given in Table 2.1

3. The H_{ij}, the off-diagonal elements in the determinant, are estimated using an approximation originally proposed by Mulliken but universally known as the Wolfsberg–Helmholz approximation (Equation 2.21); K is a constant

$$H_{ij} = \frac{1}{2} KS_{ij}(H_{ii} + H_{jj}) \qquad (2.21)$$

usually set somewhere between 1.75 and 2.0. There are some other approximations for H_{ij} which are slight modifications of this formula.

Often small variations in the VSIP values included in the calculation do not affect the qualitative results of the calculation. A large amount of the structural chemistry of both main group and transition metal complexes will, as we shall see later, be based more on the general form of the molecular orbitals (usually not very sensitive to small changes in the parameters used) rather than their exact energy or composition in terms of atomic orbitals. Figure 2.8 shows the result of an EHMO calculation

Table 2.1 Parameters for Use in Extended Hückel Calculations

A. First- and Second-Row Atoms

	Principal Quantum Number	Slater Exponents		VSIP (eV)		
		s and p	d	s	p	d
H	1	1.3		−13.6		
Li	2	0.650		−5.4	−3.5	
Be	2	0.975		−10.0	−6.0	
B	2	1.300		−15.2	−8.5	
C	2	1.625		−21.4	−11.4	
N	2	1.950		−26.0	−13.4	
O	2	2.275		−32.3	−14.8	
F[a]	2	2.600		−40.0	−18.1	
Na	3	0.733		−5.1	−3.0	
Mg	3	0.950		−9.0	−4.5	
Al	3	1.167		−12.3	−6.5	
Si	3	1.383	1.383	−17.3	−9.2	−6.0
P	3	1.600	1.400	−18.6	−14.0	−7.0
S	3	1.817	1.500	−20.0	−13.3	−8.0
Cl	3	2.033	2.033	−30.0	−15.0	−9.0

[a] A value of 2.425 is often used.

B. Slater Exponents for $(n + 1)$ s, p Orbitals and for Double Zeta nd Functions of the First-Row Transition Metals[a,b]

	s	p	$d(\zeta_i)$	Coefficients $(c_i)^c$
Ti	1.075	0.675	4.55, 1.40	0.4206, 0.7839
	1.175	0.800	4.55, 1.60	0.4391, 0.7397
V	1.200	0.750	4.75, 1.50	0.456, 0.752
	1.300	0.875	4.75, 1.70	0.476, 0.706
Cr	1.325	0.825	4.95, 1.60	0.4876, 0.7205
	1.425	0.950	4.95, 1.80	0.5060, 0.6750
Mn	1.450	0.900	5.15, 1.70	0.514, 0.693
	1.550	1.025	5.15, 1.90	0.532, 0.649
Fe	1.575	0.975	5.35, 1.80	0.5366, 0.6678
	1.675	1.100	5.35, 2.00	0.5505, 0.6260

Table 2.1 *Continued*

Co	1.700	1.050	5.55, 1.90	0.555, 0.646
	1.800	1.175	5.55, 2.10	0.568, 0.606
Ni	1.825	1.125	5.75, 2.00	0.5683, 0.6292
	1.925	1.250	5.75, 2.20	0.5817, 0.5800
Cu	1.950	1.200	5.95, 2.10	0.58, 0.62
	2.050	1.325	5.95, 2.30	0.5933, 0.5744

[a] For each metal the first row gives the parameters for the neutral atom, the second row the parameters for the $+1$ ion.
[b] From Ref. 177.
[c] $\psi_d = c_1 \phi(\zeta_1) + c_2 \phi(\zeta_2)$.

on the HF molecule where $2s$ and $2p$ orbitals are included on the F atom and the parameters are taken from Table 2.1. Note that the calculation automatically includes the hybridization or mixing between fluorine $2s$ and $2p$ orbitals.

2.5 The Angular Overlap Method

If we take the Hückel determinant (Equation 2.16) and allow the two orbitals different values of H_{ii}, then the roots of the secular determinant

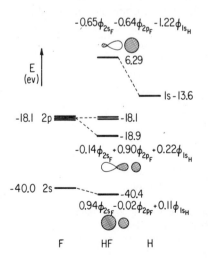

Figure 2.8 EHMO calculation on HF using a bond length of 0.9171 Å and parameters taken from Table 2.1. The larger-than-unity coefficient of the H s orbital in the highest-energy orbital arises as a consequence of the inclusion of overlap in the normalization procedure (see Section 1.3). The (small) amount of F $2p$ character in the lowest-energy orbital appears with the opposite sign to that expected. For a discussion of this molecular orbital artifact see Ref. 219.

may be written as a converging power series, assuming $H_{12}/(H_{11} - H_{22}) < 1$:

$$E_1 = H_{11} + \frac{H_{12}^2}{H_{11} - H_{22}} - \frac{H_{12}^4}{(H_{11} - H_{22})^3} + \cdots$$

$$E_2 = \overset{\circ}{H}_{22} - \frac{H_{12}^2}{H_{11} - H_{22}} + \frac{H_{12}^4}{(H_{11} - H_{22})^3} + \cdots \tag{2.22}$$

where $\psi_{1,2}$ are the bonding and antibonding orbitals, respectively ($|H_{11}| > |H_{22}|$). In this zero overlap approximation (Figure 2.9a), the energy shifts of upper and lower orbitals are equal and opposite. The orbital interaction increases the closer the two atomic orbitals are in energy ($H_{11} - H_{12}$). More exactly, overlap may be included and the secular determinant (Equation 2.23) solved. This does not give a particularly simple algebraic result. However, assuming that the energy shifts are small, by approximating the off-diagonal terms as $H_{ij} - H_{ii} S_{ij}$ (i.e., $E_i \sim H_{ii}$), Equations 2.24 result:

$$\begin{vmatrix} H_{11} - E & H_{12} - ES_{12} \\ H_{12} - ES_{12} & H_{22} - E \end{vmatrix} = 0 \tag{2.23}$$

$$E_1 = H_{11} + \frac{(H_{12} - S_{12}H_{11})^2}{H_{11} - H_{22}} \qquad E_2 = H_{22} + \frac{(H_{12} - S_{12}H_{22})^2}{H_{22} - H_{11}} \tag{2.24}$$

Of particular interest is what happens to the E_i if the Wolfsberg–Helmholz equation (2.21) is used for H_{ij} with $K = 2$. The outcome is a particularly simple form for the E_i (Equations 2.25),

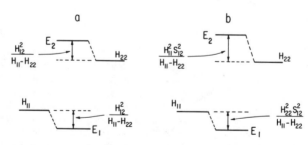

Figure 2.9 Interaction diagram for two orbitals of different energy (*a*) without overlap and (*b*) with overlap included via a perturbation method. Only the lead term in Equation 2.22 is shown.

$$E_1 = H_{11} + \frac{H_{22}^2}{H_{11} - H_{22}} S_{12}^2$$

$$E_2 = H_{22} - \frac{H_{11}^2}{H_{11} - H_{22}} S_{12}^2$$

(2.25)

shown schematically in Figure 2.9b. Note that the bonding orbital is stabilized less than the antibonding orbital is destablized, a consequence of the inclusion of overlap and arising simply because $|H_{11}| > |H_{22}|$. Thus, as two orbitals get closer together in energy, or the overlap integral between them increases, their mutual interaction energy (ϵ) increases, that is, $\epsilon \alpha S^2/\Delta E$. An exactly analogous result arises by use of the second-order perturbation expression of Equation 1.15 if we write the perturbation $\int \psi_m^{(0)} \mathcal{H}' \psi_n^{(0)} \, d\tau$ as equal either to $H_{mn} - H_{mm} S_{mn}$ or as directly proportional to S_{nm}. The stabilization energy of the bonding component increases with increasing VSIP (increasing electronegativity) of the upper of the two orbitals before interaction. The stabilization energies of Equation 2.25 should really be written in terms of a power series as in Equations 2.22. In the present case this leads to

$$\epsilon_{stab} = \frac{H_{22}^2 S_{12}^2}{H_{11} - H_{22}} - \frac{H_{22}^4 S_{12}^4}{(H_{11} - H_{22})^3} + \cdots$$

$$\epsilon_{destab} = \frac{H_{11}^2 S_{12}^2}{H_{11} - H_{22}} - \frac{H_{11}^4 S_{12}^4}{(H_{11} - H_{22})^3} + \cdots$$

(2.26)

The angular overlap model takes its name and gains its power by representing the overlap integral as a simple product of radial and angular terms $S = S_\lambda F$ as in Equation 1.21, such that Equations 2.27 result:

$$\epsilon_{stab} = e_\lambda F^2 - f_\lambda F^4$$

(2.27)

$$= \beta_\lambda S_\lambda^2 F^2 - \gamma_\lambda S_\lambda^4 F^4$$

We show two labeling schemes to be found[24,115,185] in the chemical literature.* β_λ is not to be confused with the Hückel β parameter (although they are clearly related). The parameters are defined in Equations 2.28:

$$\beta_\lambda = \frac{H_{22}^2}{H_{11} - H_{22}} \qquad \gamma_\lambda = \frac{H_{22}^4}{(H_{11} - H_{22})^3}$$

(2.28)

$$e_\lambda = \frac{H_{22}^2}{H_{11} - H_{22}} S_{12}^2 = \beta_\lambda S_{12}^2 \qquad f_\lambda = \frac{H_{22}^4}{(H_{11} - H_{22})^3} S_{12}^4 = \gamma_\lambda S_{12}^4$$

* f_λ has been invented by the author for discussion in this book.

As an example we take the interaction between the $2p_z$ and $2p_y$ orbitals of the oxygen atom in H_2O with the two hydrogen $1s$ orbital combinations of the correct symmetry. The two $1s$ orbitals are ϕ_1 and ϕ_2 so that the a_1 ligand combination to interact with p_z is given by $2^{-1/2}(\phi_1 + \phi_2)$ and the b_2 combination to interact with p_y is given by $2^{-1/2}(\phi_1 - \phi_2)$ (Figure 2.10) where we have neglected the overlap between ϕ_1 and ϕ_2 in the normalization procedure. The overlap integrals are of the form of Figure 1.1a, and we readily find

$$S(a_1) = 2^{1/2}S_\sigma \cos \theta$$
$$S(b_2) = 2^{1/2}S_\sigma \sin \theta$$

$$(2.29)$$

which leads to

$$\epsilon(a_1) = 2e_\sigma \cos^2 \theta - 4f_\sigma \cos^4\theta \qquad (2.30)$$
$$= 2\beta_\sigma S_\sigma^2 \cos^2 \theta - 4\gamma_\sigma S_\sigma^4 \cos^4 \theta$$

and

$$\epsilon(b_2) = 2e_\sigma \sin^2 \theta - 4f_\sigma \sin^4 \theta \qquad (2.31)$$
$$= 2\beta_\sigma S_\sigma^2 \sin^2 \theta - 4\gamma_\sigma S_\sigma^4 \sin^4 \theta$$

Since the hydrogen $1s$ orbital has a lower ionization potential than the oxygen $2p$ orbital, the molecular orbital energy diagram describing this interaction will take the form of Figure 2.10 where we distinguish two different e_σ and f_σ values for bonding and antibonding orbitals. In Chapter 6 we investigate the angular dependence of these functions.

There are two sum rules[188] associated with the orbital stabilization energies which are very useful. For the present case (Figure 2.10) the

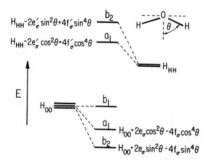

Figure 2.10 Angular overlap method energy level diagram for interaction of a_1 and b_2 symmetry orbitals in the H_2O molecule. (We use different AOM parameters, e_σ, $e_\sigma{}'$, etc., for bonding and antibonding interactions.)

sum of the coefficients of the e_σ (quadratic) part of the interaction energy for ligand interaction with p_x, p_y, and p_z (in this case there is no ligand σ combination of the correct symmetry to overlap with p_x) is equal to $2e_\sigma$, in other words, is angle independent. This is an example of a general sum rule associated with the angular part of the overlap integral, F,

$$\sum_{m_l} F^2(\lambda, l, m_l) = n_\lambda \tag{2.32}$$

where n_λ is the number of ligand λ orbitals (two in the present case). An analogous sum rule does not apply to the F^4 terms. Another sum rule that will prove useful is that often the quadratic interaction energies of the ligand orbitals with a central atom orbital are additive as a function of geometry, that is, $\epsilon_x(y$ or $z)$ for $A - Y_n = \epsilon_x(A - Y_{n1}) + \epsilon_x(A - Y_{n2})$, where $n = n_1 + n_2$. A simple example is shown in Figure 2.11. The criterion for the validity of the approach is that there are no two non-degenerate central atom orbitals with the same symmetry (or no two pairs of doubly degenerate orbitals) which may mix together in AY_n. If this criterion is met, then the interaction energy of a central atom orbital with the n ligands is given by the simple ligand additivity Equation 2.33,

$$\epsilon = \sum_n e_\lambda \cdot F_\lambda^2(n) \tag{2.33}$$

where $F_\lambda(n)$ is the angularly dependent part of the overlap integral between the central atom orbital and the nth ligand orbital. This is the extreme of the geometry sum rule where $A - Y_{n_1}$, $A - Y_{n_2}$, and so on, are diatomic units consisting of a central atom and one ligand as in fact in Figure 2.11. This result comes directly from the pairwise additivity result of second-order perturbation theory discussed in Chapter 1.

We use the AOM to look at the allene problem of Section 2.3. In the staggered arrangement we need to consider the two components of the

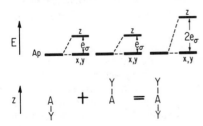

Figure 2.11 Simple illustration of the ligand additivity rule for quadratic angular overlap energies. Generation of a diagram for a linear AY_2 molecule from those for two AY species.

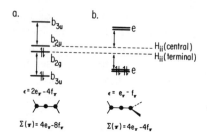

Figure 2.12 AOM diagrams for the allene molecule in (a) D_{2h} eclipsed and (b) D_{2d} staggered configurations.

e species interaction (Figure 2.12a). Clearly if the overlap integral within each pair is S_{π}, then $\epsilon = e_{\pi} - f_{\pi}$. For the eclipsed arrangement there are two ligand symmetry combinations (Figure 2.12b) of which only the one of species b_{3u} finds a central atom counterpart. The overlap integral of this function with the central atom is simply $2/\sqrt{2}S_{\pi}$, leading to $\epsilon = 2e_{\pi} - 4f_{\pi}$. With a total of four electrons the staggered form has a π molecular orbital stabilization energy (which we write as $\Sigma(\pi)$) of $4e_{\pi} - 4f_{\pi}$. For the eclipsed arrangement $\Sigma(\pi) = 4e_{\pi} - 8f_{\pi}$ which represents a smaller stabilization and is of higher energy. This approach illustrates a molecular orbital and structural result which is of widespread occurrence. If possible two ligands will avoid sharing the same central atom orbital, or if this is unavoidable the shared arrangement is of higher energy and the bonds will be weakened. While the e_{π} term (containing terms in S_{π}^2) appearing in $\Sigma(\pi)$ is independent of angle, the f_{π} term (containing terms in S_{π}^4) is not. The smallest quartic term is obviously found for the geometry where the latent quadratic energy is distributed among the largest number of interactions. The result is a general one* for those systems where the bonding orbitals from this interaction are filled (the e species orbitals in Figure 2.12a). We could use an analogous argument to rationalize the tetrahedral arrangement of ligands in the methane molecule using a p-orbital-only model. Here all three p orbitals are equally used by the ligands in σ bonding, and this geometry, with a σ molecular orbital stabilization energy $\Sigma(\sigma) = 8e_{\sigma} - (32/3f_{\sigma})$, is favored over a square planar geometry (for example) where only two in-plane orbitals are involved in σCH interactions ($\Sigma(\sigma) = 8e_{\sigma} - 16f_{\sigma}$). Exactly the opposite structural prediction

---

* If $\Sigma_{k=1}^{m} n_k = c$ (constant), then from the Cauchy–Schwartz inequality, $c^2 \geq \Sigma n_k^2 \geq c^2/m$. The minimum value of $\Sigma_{k=1}^{m} n_k^2$ is for $n_1 = n_2' = \ldots n_m$ and the maximum for $n_1 = c$ and all other $n_k = 0$. In the present case we identify n_k with S_{ij}^2, and c arises through the sum rule of Equation 2.32. Then the maximum value of the quartic term appears at the planar allene geometry where all the latent interaction is contained in one bonding orbital ($n_1 \neq 0, n_2 = 0$) and the minimum value at the twisted geometry where the interaction is shared between two orbitals ($n_1 = n_2$).

occurs for systems where all the antibonding orbitals are occupied as well. The eclipsed allene structure should be found for the eight-electron case since now the minimum *destabilization* of the antibonding orbitals occurs at this geometry. Experimentally this is found for the $HgPh_2$ molecule, for example.

Our philosophy in using the AOM will then be as follows. The energy $\Sigma(\sigma)$ or $\Sigma(\pi)$ is described by the sum of quadratic (e) and quartic (f) terms in the overlap integral. The quadratic term is the more energetic of the two. If a structural result can be resolved by considering the differences in the quadratic term for various geometric alternatives, we will ignore the less important quartic contribution. (In the analysis of the electronic spectra of transition metal complexes using the AOM, only the e_σ terms are used.) On the other hand, we often find that the quadratic term is the same for both structural alternatives, as in the allene case above. We then need to resort to consideration of the quartic terms. Two different ways have been used to calculate the quadratic part of the interaction, ϵ, of a pair of orbitals, the ligand additivity Equation 2.33, and the overlap of a symmetry-adapted ligand combination with a central atom orbital. The two methods are of course equivalent. Once the coefficient (p) of the e_λ term has been obtained, that of the f_λ term is simply p^2 leading to a general expression for the interaction energy between two orbital sets of the same symmetry as

$$\epsilon = pe_\lambda - p^2 f_\lambda \qquad (2.34)$$

We develop the AOM methodology further in other places in the book.

2.6   Reliability of the Results

All three methods provide some sort of "orbital energy." Its variation, either as a function of geometry or from one system to another, may readily be followed. These energies may be obtained either numerically using the extended Hückel method (EHMO) or algebraically in para-metrized form using the angular overlap method (AOM) or Hückel method (HMO) depending on our needs. The EHMO method has the advantage that hybridization of orbitals is automatically taken care of, as in the simple HF case, by relevant overlap integrals. Such orbital mixing is much more difficult to achieve using the AOM since we are interested here in the interaction between pairs rather than trios of orbitals. However the AOM has the advantage that for simple systems a molecular orbital diagram may be rapidly derived and in favorable cases trends in orbital energies on distortion may be noted, but it operates at a less sophisticated level than the EHMO scheme.

How reliable are the results, for example, in the prediction of geometries, using these methods? As far as angular geometries, barriers to rotation, and so on, are concerned, the answer is usually very good indeed using the EHMO method. Many of the structural features of both main group and transition metal complexes are often mirrored by the results of these simple EHMO calculations.[94,141] Many other less quantitative aspects of molecular structure[24,27] may be reached by using the ideas of the AOM. As far as matching bond lengths or predicting photoelectron or electronic spectra quantitatively, the answer is usually very poor indeed. The reason for the good fit for angular geometries probably lies in the fact that the angular variation of overlap integral is exactly given by Equation 1.21, irrespective of the chemical nature of the atoms, and is essentially determined by group theory applied to atomic systems, namely the $Y_{lm}(\theta, \phi)$.

The distance variation in overlap integral depends however on the choice of the radial part of the orbital wavefunction. This is not well defined, and Slater orbitals are perhaps rather crude estimates. A more probable reason for the failure of the EHMO method to predict bond lengths is that nuclear–nuclear repulsions are left out of the calculation completely so there is no force dissuading the nuclei from getting too close. The H_2 molecule with two $1s$ orbitals provides a dramatic example. The lowest value of the energy of the system within the extended Hückel framework occurs when the overlap integral between the two orbitals is maximized. This is for $S = 1$ where the two orbitals are superimposed, that is, when the bond length is zero! There is however a molecular orbital effect which stops nonbonded atoms from getting too close, as for example in the energy minimization method for calculating angular geometries. Each atom sees a nonbonded neighbor as a closed shell of electrons, just as two He atoms coming together each see a filled $1s$ orbital on the other atom. Figure 2.13 shows that the overall effect on the energy as these filled orbitals begin to overlap is a destabilizing (repulsive) one and in-

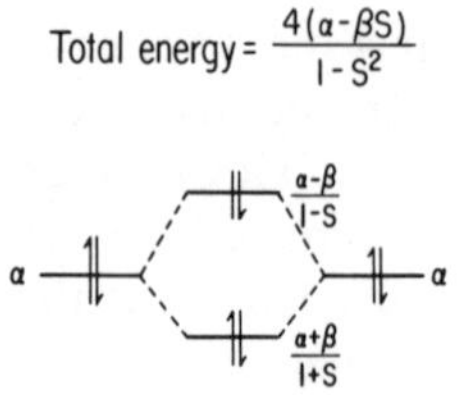

Total energy $= \dfrac{4(\alpha - \beta S)}{1 - S^2}$

Figure 2.13 Molecular orbital origin of nonbonded repulsions between closed shells of electrons.

creasingly so as the overlap integral increases. (This arises simply because the antibonding orbital is destabilized more than the bonding orbital is stabilized.) The general fact that molecular geometry is usually predicted far better than spectroscopic or other molecular properties using these crude approaches may well be due to the observation that the configurational potential depends only to second order on the electronic wavefunction.[64] Many other properties are much more sensitive to the quality of the molecular wavefunction used in their evaluation.

3 The Shapes of Main Group Molecules and the Electron Pair Repulsion Approach

In this and the following chapters we are interested in the relationship between the molecular electronic configuration, the nature of the chemical bonds, and the overall structure and geometry of the molecule. Most of our effort will be devoted to investigating the shape or angular geometry of these species and associated variations in bond lengths. (We note here that the structures of some systems may be phase dependent.[87]) There is a diversity of models used to view main group stereochemistry and electronic structure.[42,59,73] At first sight they may seem to have little common ground but we shall show that the molecular orbital-based models at least are related.

3.1 The Valence Shell Electron Pair Repulsion (VSEPR) Model

It has been suggested[71,72] that the most favorable packing of pairs of electrons in the valence shell of a central atom dictates the bond angles and relative bond lengths in molecules. Over the years a very useful approach for predicting the structures of simple molecules has been developed by Sidgwick, Powell, Nyholm, and Gillespie—the VSEPR method. It may be presented as four basic rules of which the first two are sufficient to determine the gross geometry.

1. The best arrangement of a given number of electron pairs in the valence shell of an atom is that which maximizes the distances between them.

2. A nonbonding pair of electrons occupies more space on the "surface" of an atom than a bonding pair.
3. The "size" of a bonding pair decreases with increasing electronegativity of the ligand it connects to the central atom.
4. The two-electron pairs of a double bond or three-electron pairs of a triple bond occupy more space than the one-electron pair of a single bond.

Rule 1.   Each pair of electrons is assumed to occupy a well-defined region of space around the central atom, and all other pairs of electrons are excluded from this space by "Pauli repulsion" or "Pauli avoidance" forces.[13] While electrons of opposite spin only are allowed to occupy the same region of space, electrons with the same spin must keep apart. Since there are only two values of the spin quantum number, this results in electron *pairs* occupying a reasonably well-defined region of space and "repelling" other electron pairs. We note here that most of the objections leveled at this model of molecular structure have been concerned with its lack of physical reality.[51,197] In quantum mechanical language, this force is the expectation value of an operator. It is however difficult to formulate this operator mathematically, which makes the estimation of the relevance of these forces somewhat difficult. We are not in fact able at present to convincingly relate the VSEPR and MO models to each other. We may see which geometries give rise to the lowest energy structures from this rule by calculating the minimum energy arrangement for n-like charges (modeling the n electron pairs) constrained to move on the surface of a sphere subject to a given mutual interaction potential, with the central atom and its core of electrons sitting in the middle. This gives rise to the geometric arrangements of charges or electron pairs shown in Table 3.1

Table 3.1   Predicted Arrangements of n Electron Pairs Using the VSEPR Approach

Two	Linear
Three	Equilateral triangle
Four	Tetrahedron
Five	Trigonal bipyramid
Six	Octahedron
Seven	Monocapped octahedron
Eight	Square antiprism
Nine	Tricapped trigonal prism
Ten	Bicapped square antiprism
Eleven	Icosahedron minus one apex
Twelve	Icosahedron

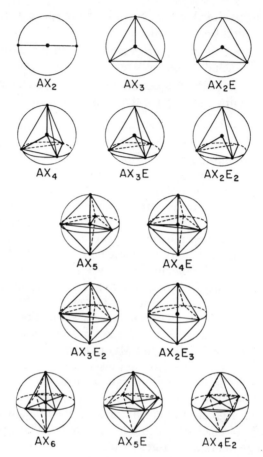

Figure 3.1   Minimum energy arrangements of n points on a sphere to simulate arrangement of electron pairs around a central atom. Reprinted with permission from *Molecular Geometry* by R. J. Gillespie, Van Nostrand-Rheinhold, London (1972). Copyright by R. J. Gillespie.

and Figure 3.1. Thus two charges clearly are oriented diagonally opposite each other on the sphere for lowest energy leading to a linear arrangement of the electron pairs relative to the central atom. Similarly three charges locate themselves in an equilateral triangle about the central atom. For $n \geq 7$ the results are not intuitively obvious (perhaps for $n = 5$ also), and in these cases the actual result depends upon the details of the potential invoked between the charges. As we shall see, systems with such large numbers of electrons are usually flexible, indicating that several arrangements are close in energy.

The electrons to be included when counting up the number of electron pairs around the central atom are those ligand electrons involved in σ interactions with the central atom plus all this atom's valence electrons. In most cases one electron is contributed per ligand if the ligands are counted as uncharged entities. Tetrahedral CF_4 contains eight valence electrons (four pairs), four from the central carbon atom $(2s^2 2p^2)$ and one each from four fluorine atoms. In some systems however it is best to regard the ligands as contributing two or zero σ electrons. For $BF_3 \cdot NH_3$, the four electron pairs surrounding the B atom arise from the three boron electrons $(2s^2 2p^1)$, one electron from each of the fluorine atoms, and two electrons from the lone-pair orbital on the nitrogen atom. Similarly the four pairs around the N atom arise from the five nitrogen electrons $(2s^2 2p^3)$, one electron from each of the three hydrogen atoms and none from the coordinated boron since the p_z orbital directed toward the nitrogen atom is empty.

In the case of two, three, and four electron pairs, rule 1 leads to no ambiguity. $BeCl_2$ and $CaCl_2$ as free gas-phase molecules are linear (two pairs of electrons, AB_2). CH_2 is a bent species in singlet and triplet states, but in the singlet with three pairs a bent AB_2E geometry is found (B represents a bonded electron pair, E an unshared pair.) OH_2 with four pairs is also bent (AB_2E_2); the two shared pairs occupy two corners of a tetrahedron on this scheme, and the two unshared pairs occupy the other two. Similarly BF_3 is trigonal planar (AB_3), but NF_3 is pyramidal (AB_3E). One rather unusual molecule which we may also include in the scheme is the $(PPh_3) \cdot C \cdot (PPh_3)$ species (or the isoelectronic $(PPh_3) \cdot C \cdot CO$). Its bent PCP skeleton is in accord with a total of four valence pairs, AB_2E_2. Four electrons come from the carbon atom and two each from the PPh_3 units just as for the isoelectronic NH_3 molecule in the $BF_3 \cdot NH_3$ complex above. Often molecules with an odd electron (a half-pair) do not provide any special problems. Thus NF_2 (three and a half pairs) is a bent molecule commensurate with the bent species CF_2 (three pairs) and OF_2 (four pairs). BH_2 (two and a half pairs) is however bent and lies between $BeCl_2$ (two pairs and linear) and CF_2. CH_3 contrarily is planar (three and a half pairs), whereas BF_3 (three pairs) is planar and NF_3 (four pairs) is pyramidal. How much "space" half a pair occupies clearly is system dependent.

Rule 2. For a molecule with five or six electron pairs we need also to consider this rule to define the geometry. The idea here is that a shared pair of electrons under the influence of two nuclei will be less free to move than an unshared pair influenced by a single nucleus. So repulsions involving bonded pairs are smaller than repulsions involving nonbonded pairs (i.e., bonded–bonded < bonded–nonbonded < nonbonded–nonbonded). Thus unshared electron pairs prefer the equatorial rather than axial sites

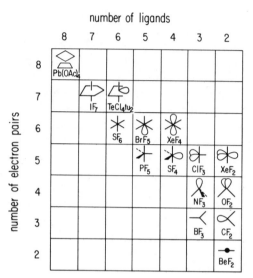

Figure 3.2 Range of structures adopted by molecules governed by the VSEPR rules 1 and 2 (tu = thiourea).

in a trigonal bipyramid (five pairs). Figure 3.2 shows the limitations on the range of structures adopted as a result of this rule along with a tabulation of predicted molecular structures as a function of the number of bonded and unshared pairs of electrons. In general the scheme works remarkably well. There are some exceptions to the predictions. Some five-pair molecules, $SbPh_5$, SbF_5, and perhaps $Sb(C_3H_5)_5$, exhibit square pyramidal rather than trigonal bipyramidal geometries although as Figure 3.3 shows the transition from one to the other requires relatively small atomic shifts, and a "spectrum" of structures is actually observed.[145] The linear XeF_2 geometry, with six pairs, is again consistent with rule 2. (In the *trans*-octahedral geometry that is observed, each unshared pair has two unshared pairs as 90° neighbors, in the alternative *cis*-octahedral structure two unshared pairs have two unshared pairs as 90° neighbors, and two have three.)

One consequence of rule 2 that shows up in the structural results is the sizes of bond angles and bond lengths not set by symmetry requirements. Thus in NY_3 species the YNY angles are less than tetrahedral since on this scheme the bulky lone pair repels the bonding pairs more than the mutual repulsions between them. Similarly ClF_3 is distorted toward the arrowhead geometry rather than to a Y shape from the T shape geometry, and in square pyramidal BrF_5 the axial/basal angle is less than 90° giving rise to an "umbrella" geometry. In SF_4 the two axial bonds

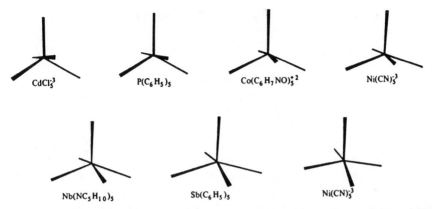

Figure 3.3 Transition from trigonal bipyramidal geometries (top left) to square pyramidal (bottom right). Reprinted with permission from E. L. Muetterties and L. J. Guggenberger, *J. Am. Chem. Soc.* **96** 1751 (1974). Copyright by the American Chemical Society.

are tilted toward the two equatorial ones ("external" FSF angle 187°). In AY_4-substituted species with this geometry, for example, $TeBr_2(C_6H_5)_2$, $SeCl_2(C_6H_4CH_3)$, the axial bonds tilt in the opposite sense (bond angles 178° and 177.5°, respectively). This may be a steric effect. $TeCl_2Me_2$ has a ClTeCl angle of 188°. Another facet of this rule is that bonds adjacent to lone pairs experience the largest repulsions and will be longer than those further away. This is shown convincingly by the structures of ClF_3, SF_4, and BrF_5. A collection of structural results of importance for these molecules is shown in Figure 3.4. Note that in each case two long bonds are trans to each other, but a short bond is trans to a vacancy.

Examples with seven, eight, and nine pairs of electrons are rare for main group systems. For IF_7 (AB_7) a pentagonal bipyramidal geometry is found. The molecule is fluxional,[3] and the five equatorial atoms execute a puckering motion best described as a pseudorotation. The molecule

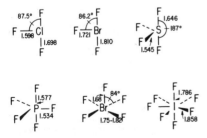

Figure 3.4 Some structural results of interest (bond lengths in Å).

ReF$_7$($d°$) behaves similarly.[103] For XeF$_6$ (AB$_6$E) the geometry of the gas-phase molecule is nonoctahedral,[68] but the species is flexible having a near-zero force constant for one of the bending modes of the octahedron. In the solid state[22] complex polymeric species are found in all crystalline modifications. For this system, in addition to the VSEPR forces, the non-bonded repulsions of the six fluorine atoms are also important. For IF$_7$ these repulsions must also be taken into account, but the geometric preferences of the seven electron pairs will be similar to those of the seven nonbonded atoms. Other experimental results on the geometries of AB$_6$E species provide a variety of results. Ions such as AX$_6$$^{-2}$ (A = Se or Te, X = Cl or Br) are regular octahedral species in the crystal.[129,215] The vibrational spectra of the TeX$_6$$^{-2}$ ions however are interpreted[2] in terms of large amplitudes and abnormal frequencies, suggesting that this geometry is perhaps rather tenuously stable. The SbX$_6$$^{-3}$ species seem to be distorted in the crystal toward a C_{3v} geometry where the extra electron pair sticks out of one face of the octahedron (capped octahedron of electron pairs). The electron diffraction data for gas-phase XeF$_6$ are explicable also on the basis of a C_{3v} structure with a C_{2v} geometry (Figure 3.5) and an octahedral arrangement lying higher in energy. One form of *trans*-TeCl$_4$ (thiourea)$_2$ has a C_{2v} skeletal geometry in the crystal with the lone pair pointing between two ligands giving a pentagonal bipyramidal arrangement of electron pairs.[57] Another form is octahedral. Of course all that can be determined experimentally are the atom positions. We infer the position of the lone pair from the nature of the geometric distortion. The fascinating problem of these AB$_6$E molecules is discussed at length in later chapters.

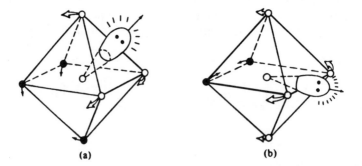

(a)   (b)

Figure 3.5 Possible "positions" of the lone pair of electrons in XeF$_6$ consistent with the electron diffraction study and the resultant distortion of the octahedral geometry. Reprinted with permission from L. S. Bartell and R. M. Gavin, *J. Chem. Phys.* **48** 2470 (1968).

Rule 3. As the electronegativity of a ligand increases, the electron pair shared by the ligand will be more contracted and tightly bound than an electron pair tied to a less electronegative atom. Thus the "space" occupied by a shared pair of electrons should decrease as the ligand electronegativity increases. So OF_2 and NF_3 have smaller bond angles than OH_2 and NH_3. As a rough rule the BAB bond angles increase in the order $F < Cl < Br \sim I$ for AB_2E_2 and AB_3E species (Tables 3.2 and 3.3). The hydrides of the heavier elements are exceptions. The HAH angles in species such as PH_3, AsH_3, and SH_2 are close to 90° and smaller than those for any corresponding halide. They are usually regarded as being anomalous on the VSEPR scheme.

Rule 4. This allows explanation of the fact that angles containing multiple bonds are usually larger than those involving single bonds. Some examples are shown in Table 3.4.

There are four main areas where the general ideas of the VSEPR scheme fail, although there are several places where unusual geometries may be forced on molecules by suitable ligand choice. One example might be P(III) porphyrin where a square-planar geometry is forced upon this AB_4E molecule by the structure of the ligand:

a.   Where there are obviously polar bonds. Thus Li_2O is linear and not bent as its AB_2E_2 structure predicts. If the molecule is regarded

Table 3.2   Geometric Data for AY_2 and AH_2 Species—Bond Lengths (Å) and Angles (°)

	O	S	Se	Te	Be	Mg	Ca	Sr	Ba[a]
AH_2	0.957	1.33	1.46						
	104.9	92.2	91	89.5					
AF_2	1.413	1.59							
	103.8	98			180	159	140	108	~100
ACl_2	1.70	2.00							
	110.8	103			180	180	180	119	~100
ABr_2				2.51					
	111			98	180	180	180	180	[b]
AI_2									
					180	180	180	180	[b]
$A(CH_3)_2$	1.416	1.80	1.98						
	111.5	99	98						

[a] AX_2 species for A = Pb, Sn are nonlinear; for A = Zn, Cd, Hg, the molecules are linear.
[b] Nonlinear.

Table 3.3 Geometric Data for AY_3 and AH_3 Species—Bond Lengths (Å) and Bond Angles (°)

	N	P	As	Sb	Bi
AH_3	1.015	1.42	1.519	1.71	
	106.6	93.5	91.8	91.3	
AF_3	1.371	1.54	1.71	2.03	
	102.1	100	102		
ACl_3	1.759	2.04	2.16	2.33	2.48
	107.1	100.1	98.4	99.5	100
ABr_3		2.18	2.33	2.51	2.63
		101.5	100.5	97	100
AI_3		2.43	2.55	2.67	
		102	101.5	99	
$A(CH_3)_3$	1.47	1.84	1.98		
	108	98.9	96		

Table 3.4 Multiple Bonds and Bond Angles in $AY_2 = Y'$ Systems

	YAY Angle (°)	Number of σ Pairs and Angle Expected on σ-Only Model	
$CF_2 = O$	112.5	(3) AB_3	(120)
$CH_2 = O$	118	(3) AB_3	
$SF_2 = O$	92.8	(4) AB_3E	(109.5)
$SeCl_2 = O$	106	(4) AB_3E	
$PF_3 = O$	102.5	(4) AB_4	(109.5)
$PF_3 = S$	100.3	(4) AB_4	
$O = SF_2 = O$	96.1	(4) AB_4	

as being made up as $Li^+O^{-2}Li^+$, then clearly the best way of arranging the two Li^+ ions is diagonally around the O^{-2} ion.

b. Molecules with π systems. $C(CN)_3^-$, $C(NO_3)_3^-$, and several other AB_3E species predicted to be pyramidal are found to have planar skeletons. One explanation is that π bonding stabilizes the planar structure.

c. "Inert pair" molecules. The ns^2 pair of electrons in the heavy elements of the right-hand side of the periodic table is often found to be stereochemically inert. As we have seen $SbCl_6^{-3}$ and $BiCl_6^{-3}$ are octahedral species but with seven valence pairs (AB_6E). XeF_8^{-2} (as the

Cs salt), which has nine electron pairs, is found as the square anti-prismatic geometry (predicted for eight pairs and found experimentally for lead tetraacetate). In each case the n electron pair molecule exhibits the structure expected for the $n - 1$ electron pair species.

d. Another area where the VSEPR scheme does not work is in viewing molecular conformations. Thus the lowest energy conformation of $H_2H{-}NH_2$ is the one where the "lone pairs" are in the gauche (Figure 3.6) rather than in the anti arrangement predicted by minimizing lone pair–lone pair repulsions on the VSEPR approach. Many molecules of this type exhibit this "gauche effect," which may be defined as "a tendency to adopt that structure which has the maximum number of gauche interactions between adjacent electron pairs and or polar bonds."[222] The observed conformations may be fitted into a new scheme by changing the VSEPR rules for the sizes of interactions between an electron pair on one atom and a polar bonded pair or lone pair on an adjacent atom.

3.2 The Directed Valence Approach[159]

The directions the bonds point in space are set in this model by the nature of the hybrid orbitals formed on the central atom to hold n electron pairs by using a valence bond approach.[119] Table 3.5 gives the angular disposition of various hybrids synthesized by relevant combinations of s, p, and d orbitals. Since the energies of the orbitals increase in this order, the lowest energy hybrids will contain the largest possible amount of s and p character and the smallest amount of d. There is no problem or ambiguity about systems with four pairs or less. Two pairs are housed in the diagonal lobes of two sp hybrids, three pairs in the trigonal sp^2 hybrids, and four pairs in tetrahedrally located sp^3 hybrids. The geometries adopted by AB_2, AB_3, AB_2E, AB_4, AB_3E, and AB_2E_2 species then readily follow since the bond angles are set by the hybrid orbital directions which are identical to those on the VSEPR scheme. When $n \geq 5$ a problem arises, since the hybrids must be of the sp^3d^{n-4} type; and for $n = 5$ two possibilities arise (Table 3.5). For $sp^3d_{z^2}$, a trigonal bipyramid (tbp)

 N_2H_4, P_2H_4

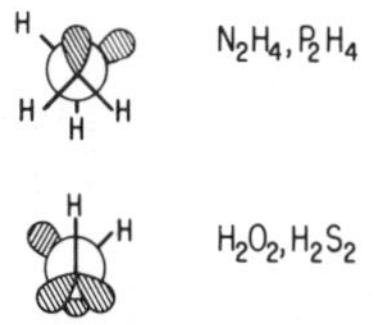

 H_2O_2, H_2S_2

Figure 3.6 Orientation of lone pairs illustrating the gauche effect.

Table 3.5 Directional Hybrid Orbitals

Number of Hybrids	Type	Spatial Arrangement
2	sp	Linear
	dp	Linear
	p^2	Angular
	ds	Angular
	d^2	Angular
3	sp^2	Trigonal plane
	dp^2	Trigonal plane
	ds^2	Trigonal plane
	d^3	Trigonal plane
	dsp	Unsymmetrical plane
	p^3	Trigonal pyramid
	d^2p	Trigonal pyramid
4	sp^3	Tetrahedron
	d^3s	Tetrahedron
	dsp	Trigonal plane
	d^2p^2	Trigonal plane
	d^2sp	Distorted tetrahedron
	dp^3	Distorted tetrahedron
	d^3p	Distorted tetrahedron
	d^4	Trigonal pyramid
5	dsp^3	Trigonal bipyramid
	dsp^3	Square pyramid
	d^3sp	Trigonal bipyramid
	d^2sp^2	Trigonal pyramid
	d^4s	Square pyramid
	d^2p^3	Square pyramid
	d^4p	Square pyramid
	d^3p^2	Pentagonal plane
	d^5	Pentagonal pyramid
6	d^2sp^3	Octahedron
	d^4sp	Trigonal prism
	d^5p	Trigonal prism
	d^3p^3	Trigonal antiprism

Table 3.5 (*Continued*)

	d^3sp^2	Mixed[a]
	d^5s	Mixed
	d^4p^2	Mixed
7	d^3sp^3	Capped octahedron
	d^5sp	Capped octahedron
	d^4sp^2	Capped trigonal prism[b]
	d^4p^3	Capped trigonal prism
	d^5p^2	Capped trigonal prism
8	d^4sp^3	Dodecahedron
	d^5p^3	Square antiprism
	d^5sp^2	Face-centered prism

Adapted from Ref. 119.
[a] Low-symmetry structure
[b] Capped on square face

is predicted, whereas a square-based pyramid (spy) is predicted for $sp^3d_{x^2-y^2}$. It is assumed by consideration of the experimental data that the former hybridization gives rise to stronger bonds and that use of the $sp^3d_{z^2}$ hybrids will be appropriate here. It is also assumed that the hybrids containing an unshared electron pair prefer the equatorial positions of the tbp just as in the VSEPR model of the previous section. Given these two assumptions, the predictive methodology is then very much like that of the VSEPR scheme. The only difference between the two approaches lies in the physical factors invoked to control the spatial arrangement of electron pairs. In the VSEPR scheme it is the Pauli avoidance of electron pairs, whereas in the present approach it is the angular directions of the lowest-energy hybrid orbitals. One still needs however to include rules 2 through 4 of the VSEPR scheme to rationalize finer points of the geometries of some of these systems, for example, the bond angles and bond lengths in ClF_3. The weakest part of the model is the insistence on including central-atom d orbitals on an equal footing with the s and p orbitals. In the molecular orbital approach to main group structures which follows, a sensible structural picture is obtained without their inclusion.

4 Molecular Orbital Models of Main Group Stereochemistry: The Walsh Diagrams

Walsh estimated[213] in qualitative terms, by building on earlier ideas of Mulliken, the energy changes associated with the molecular orbitals of one structure as it was deformed into another. If the orbital energy changes associated with the $D_{\infty h} \rightarrow C_{2v}$ distortion of AY_2 molecules, for example, can be estimated, then it may be possible to predict which electronic configurations will exhibit a bent geometry. (This after all is what is done numerically in quantitative molecular orbital calculations.) The generalizations concerning the shapes of various AY_n systems as a function of the number and arrangement of valence electrons are known as Walsh's rules. The term Walsh (or more rarely Walsh–Mulliken) diagram has come to describe in general a molecular orbital diagram where the orbitals in one symmetrical or reference geometry are correlated in energy with the orbitals of the deformed structure, irrespective of whether the energy changes were estimated using qualitative, semiempirical, or more sophisticated methods.[20] In this chapter we describe using qualitative ideas how the diagrams may be derived for simple systems without resorting to quantitative molecular orbital procedures (or alternatively how quantitative results may be understood). The key idea behind the Walsh approach to molecular geometry is that it is the energetic behavior of the highest occupied molecular orbital (HOMO) on distortion that determines the molecular geometry.

4.1   AH₂ and AY₂ Systems

It is a simple matter to derive a molecular orbital diagram for a linear dihydride species. Figure 4.1 shows two cases: (a) where the ionization potential of the A-atom s and p orbitals straddle that of the ligand ionization potential and (b) where the ligand ionization potential is smaller than that for either the central-atom s or p orbitals. The change in energy of these orbitals with angle is simply derived in qualitative fashion (Figure 4.2). The σ_u^+ levels of the linear structure are constructed from the in- and out-of-phase overlap of central-atom p_y orbitals with the ligand σ orbitals ($1s$ in the case of hydride ligands but in general any σ-type orbital on the ligand). As the angle θ decreases from 90°, the overlap integral between each lobe of the p orbital and the ligand σ orbital decreases (recall Equation 1.22 and Figure 1.1a). As the overlap integral decreases, the bonding orbital becomes less bonding and rises in energy, and the antibonding orbital becomes less antibonding and drops in energy. The $1\pi_u$–$1b_1$ orbital is of the wrong symmetry to interact with the ligand σ orbitals, either in the bent or linear geometry and so remains unchanged in energy. There are three a_1 symmetry orbitals in the C_{2v} molecule. The lowest, which is very largely As in character, usually changes very little in energy on distortion. A slight stabilization is depicted in Figure 4.2 to allow for some positive overlap of the two ligand σ orbitals as the geometry is changed. Walsh actually allowed the energy of this orbital to rise on distortion. We return to this point in Chapter 7.) The other two a_1 orbitals lie higher in energy, and it is on these orbitals that our attention will be concentrated. They are of different symmetry in the linear molecule (σ_g^+, π_u) but become of the same symmetry, and so will mix together,

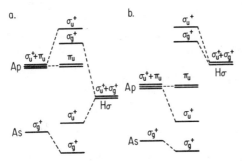

Figure 4.1   Molecular orbital diagram for a linear AH₂ molecule showing two possibilities of the values of the A-atom s and p orbital ionization potentials relative to those of H $1s$.

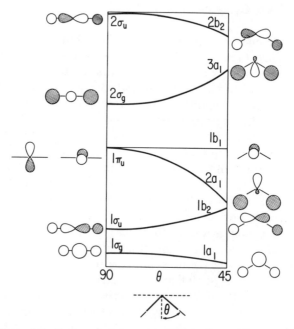

Figure 4.2 Energy changes on bending the linear AH_2 structure.

on distortion. The overall result, readily seen from the perturbation theory conclusions of Chapter 1, is that these energy levels "repel" each other and the wavefunction describing each of the new orbitals is a mixture of the old. Both $2a_1$ and $3a_1$ then contain admixtures of s and p orbitals on the central atom for any geometry other than linear as shown in Figure 4.2. An important consequence is that the lower a_1 component $(2a_1)$ is dramatically stabilized on bending.

We may see how this occurs algebraically by recalling that the numerator of Equation 1.10 is proportional to the overlap integral of the two orbitals which mix together when the perturbation is switched on (when the molecule is distorted). At the linear geometry the wavefunctions describing the $1\pi_u$–$2a_1$ and $2\sigma_g^+$–$3a_1$ orbitals are readily written down as

$$\psi(2a_1) = \phi_p$$
$$\psi(3a_1) = -a\phi_s + b(\phi_1 + \phi_2)$$

(4.1)

Here $\phi_{1,2}$ are the ligand σ orbitals; a and b are orbital coefficients, and

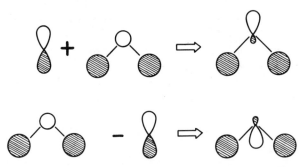

Figure 4.3 Mixing of a_1 orbitals on bending using the rules of perturbation theory.

$\phi_{s,p}$ are the As and Ap orbitals. The desired overlap integral is then

$$-a \int \phi_p \phi_s \, d\tau + b \int \phi_p(\phi_1 + \phi_2) \, d\tau = 2b \cos \theta \, S_\sigma \qquad (4.2)$$

by making use of the result of Figure 1.1. At the linear geometry (cos θ = 0) the orbitals are orthogonal (where of course they have different symmetries), but for any other geometry the overlap integral is nonzero and they will mix together, the mixing increasing on bending. Thus as the molecule bends away from the linear geometry the interaction energy between $2a_1$ and $3a_1$ is proportional to $b^2 S_\sigma^2 \cos^2 \theta$ (from Equations 1.15 and 4.2). This implies that the larger the ligand σ–central-atom p orbital overlap integral S_σ, the larger the ligand σ–central-atom s interaction (via the molecular orbital coefficient b) or the closer in energy the $2\sigma_g^+$ and $1\pi_u$ orbitals via the perturbation denominator of Equation 1.15, the larger the stabilization of $2a_1$ and destabilization of $3a_1$. In addition the $2a_1$–$3a_1$ energy separation increases with θ, and the lower-energy component (via Equation 1.10) contains an increasing amount of As character.

A similar diagram applies to AY₂ systems. There are smaller energy changes associated with the π-type orbitals than with the σ orbitals in AY₂ or AHY systems due to smaller overlap between central atom and ligand orbitals of π symmetry than those of the σ framework. Figure 4.4 shows the diagrams derived by Walsh for AY₂ systems and AHY. Assuming that this stabilization on bending of $2a_1$ is energetic enough to overwhelm the stabilization of the $1b_2$ orbital (i.e., that the curvature of the HOMO determines the geometry), then predictions as to the linearity or otherwise of these AH₂ systems may readily be made. Some examples are shown in Tables 4.1 and 4.2. With two electron pairs (e.g., BeCl₂) a linear geometry is expected; with three or four pairs a bent structure

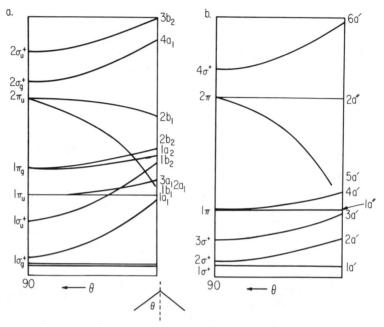

Figure 4.4 Walsh diagrams for (a) AY_2 and (b) AHY systems. The lowest two orbitals in (a) and the lowest orbital in (b) are predominantly s orbitals located on the Y ligands.

(singlet CF_2 or OF_2, respectively) is expected. With five electrons (two and a half pairs) we need to weigh the destabilizing effect of two electrons in $1b_2$ against the stabilization of one electron in $2a_1$. BH_2 and triplet CH_2 are bent molecules, although the bond angles ($\sim131°$) are not as small as in systems with two electrons in this orbital. Isoelectronic NO_2 occupies a similar place in the AY_2 series. NO_2^+ has an angle of 180°, NO_2 an angle of 134°, and NO_2^- an angle of 115.4°. Satisfyingly the linear geometry of XeF_2 or KrF_2 is nicely predicted on the approach. We get the same answers as if we had used the σ-only diagram (for XeH_2). Walsh was particularly interested in excited-state geometries, and unlike the VSEPR scheme these are readily viewed using this approach. Table 4.2 shows some predictions and observations.

4.2 AH_3 and AY_3 Systems

Figure 4.5 shows a molecular orbital diagram for a planar AH_3 system. On pyramidalization the overlap integral between the ligand σ orbitals

Table 4.1 Some Geometries of AHY and AY$_2$ Molecules

No. of Valence Electrons	Isoelectronic With	Molecule	Apex Angle (°)
10	BeH$_2$	HCN	180
		HCP	180
11	BH$_2$	HCO	119.5
12	CH$_2$	HNO	108.6
16	BeH$_2$	Numerous	180
17	BH$_2$	NO$_2$	134
18	CH$_2$	NOCl	116
		NOBr	117
		NO$_2^-$ (in crystal)	115
		O$_3$	116.5
		SO$_2$	119
19	NH$_2$	ClO$_2$	116.5
20	OH$_2$	F$_2$O	101
		Cl$_2$O	110.8
		Cl$_2$S	101
		Br$_2$Te	98
22	XeH$_2$	BrIBr$^-$	
		ClICl$^-$	
		I$_3^-$	180 or
		ClIBr$^-$	Ca. 180
		XeF$_2$	

and the p_x, p_y orbitals decreases. Thus $1e$ is destabilized and $2e$ is stabilized on bending (Figure 4.6) in an exactly analogous way to the b_2 orbitals of the AH$_2$ case. As before, the energy of $1a_1$ remains about the same on distortion, but $2a_1$ and $3a_1$ orbitals mix strongly together. They are of different symmetry species ($2a_1'$ and $1a_2''$) at the planar D_{3h} geometry but become the same (a_1) on bending. Their behavior is completely analogous to the a_1 orbitals in the AH$_2$ case. Thus BH$_3$ and BY$_3$ systems are planar, NH$_3$, NY$_3$ are pyramidal. With four electron pairs the stereochemically active lone pair in this AB$_3$E system is generated on bending the molecule (Figure 4.6); $2a_1$ is an orbital which at the pyramidal geometry contains a good admixture of s and p orbital character with a lobe pointing away from the ligands. The pyramidal NY$_3$ geometry is only predicted if the two electrons in $2a_1$ are stabilized more than the four electrons in $1e$ are destabilized on bending.

Recall a similar consideration in the AY$_2$ case. Just as BH$_2$ and triplet CH$_2$ occupied intermediate positions in the AY$_2$ series so CY$_3$ molecules

Table 4.2 Excited-State Geometries

Molecule	Bond Length^a (Å)	Angle (°)	Electronic Configuration	Comments
BH_2	1.18 1.17	131 180	$(1a_1)^2(1b_2)^2(2a_1)$ $(1a_1)^2(1b_2)^2(1b_1)$	$2a_1$ is stabilized on bending; $1b_1$ is not (see Figure 3.2)
AlH_2	1.59 1.53	119 180	As above	As above
NH_2	1.024 1.00	103.4 144	$(1a_1)^2(1b_2)^2(2a_1)^2(1b_1)$ $(1a_1)^2(1b_2)^2(2a_1)(1b_1)^2$	With only one electron in $2a_1$ excited-state molecule is less bent; cf ground-state BH_2
PH_2	1.428 1.403	91.5 123.1	As above	As above
OH_2	0.956	105.2 Nonlinear	$(1a_1)^2(1b_2)^2(2a_1)^2(1b_1)^2$ $(1a_1)^2(1b_2)^2(2a_1)^2(1b_1)(3a_1)$	With two electrons in $2a_1$, both states are bent
CF_2		Nonlinear Nonlinear		
SiF_2	— —	101.0 Nonlinear		
CH_3	1.079 1.12	120 120	$(1e')^4(1a_2'')^1$ $(1e')^4(3a_1')^1$	$3a_1'$ destabilized on bending

NH_3	1.0173 1.08	107.8 120	$(1e')^4(1a_2'')^2$ $(1e')^4(1a_2'')(3a_1')$	$5a'$ is strongly stabilized on bending [c]
HCN	1.156^a, 1.064^b 1.295, 1.14 1.334, (1.14)	180 125.0 114.5	$\cdots(1\pi)^4$ $\cdots(3a')^2(1a'')^2(4a')^2(5a')$ $\cdots(3a')(1a'')(4a')^2(2a'')$	
HCP	1.542, 1.067 1.69, (1.14) —	180 128 180	$\cdots(1\pi)^4$ $(3a')^2(1a'')^2(4a')^2(5a')$ $(3a')^2(1a'')(4a')^2(2a''),(1\pi)^3(2\pi)^1$	As above
HCO	1.19, (1.08) 1.187, 1.04	119.5 180	$(3a')^2(1a'')^2(4a')^2(5a')$ $(3a')^2(1a'')^2(4a')^2(2a''),(1\pi)^4(2\pi)^1$	As above
HNO	1.241, 1.063 1.221, 1.036	108.6 116.3	$(3a')^2(1a'')^2(4a')^2(5a')^2$ $(3a')^2(1a'')^2(4a')^2(5a')(2a'')$	$5a'$ is stabilized on bending; $2a''$ is not; ground-state angle smaller than in HCO since now two electrons in $5a'$
HCF	(1.121), 1.314 (1.121), 1.297	101.6 127.2	As HNO	As above
$HCCl^d$	1.12, 1.69 —	103.4 134	As HNO	As above

Data from Ref. 91.

[a] Ground state first, excited states second. Values in parentheses are assumed.

[b] HX last.

[c] This should have a linear geometry on the scheme cf HCP.

[d] Similar results are found for HSiCl and HSiBr.

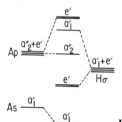

Figure 4.5   Molecular orbital diagram for a planar AH$_3$ molecule.

occupy a special place here. CH$_3$ is a planar molecule but only tenuously so. It has a low frequency out-of-plane bending mode (v_2, a_2'') which would convert the D_{3h} to the C_{3v} geometry. CH$_{3-x}$F$_x$ species are nonplanar however, the degree of bending increasing as x goes from 1 to 3. The energetics of these systems with half-filled a_1 orbitals which change dramatically in energy with geometry change are clearly rather delicately balanced on the scheme. We return to the fascinating story of CH$_3$ and related molecules in the next chapter. Some examples of AY$_3$ geometries as a function of electronic configuration are shown in Tables 4.2 and 4.3.

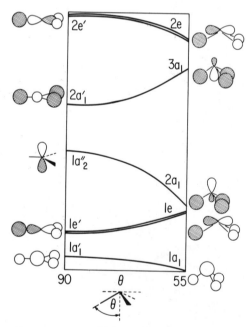

Figure 4.6   Energy changes on pyramidalization of the trigonal planar AH$_3$ structure.

Table 4.3 The Geometries of Some AY_3 Molecules

Number of Valence Electrons	Isoelectric with		Molecule		Shape
24	BH_3		CO_3^{2-}	NO_3^-	
			SO_3	NO_2Cl	
			BF_3	BO_3^{3-}	Planar
			BCl_3	InI_3	(120°)
			BBr_3	$GaCl_3$	
			$COCl_2$	$GaBr_3$	
			$COBr_2$	$CSCl_2$	
25	CH_3		ClO_3	BrO_3 (114°)	Pyramidal
26	NH_3	IO_3^-	PF_3	SbI_3	
		ClO_3^-	PCl_3	$BiCl_3$	
			PBr_3	$BiBr_3$	
			PI_3	$SOCl_2$	Pyramidal
			$AsBr_3$		(96–100°)
			AsI_3	$SOBr_2$	
			$AsCl_3$		

With five electron pairs we reach the ClF_3 or the (unknown) FH_3 molecule. This species is predicted to be resistant to pyramidalization since it has two electrons in $3a_1$. ClF_3 is indeed planar, but it has a T-shaped geometry. Three coordinate molecules have two angular degrees of freedom, in-plane bending (to a T or Y shape) and out-of-plane bending (to the pyramidal structure). Walsh only considered the latter distortion, but we may, with a little more difficulty than before, construct a qualitative energy correlation diagram for this particular deformation mode. (Alternatively we could perform a molecular orbital calculation to help us out.) Figure 4.7a shows the orbitals of the planar D_{3h} structure and how we would expect them to change in energy purely on overlap grounds without allowing any mixing to occur between the four orbitals of a_1 symmetry and retaining the D_{3h} descriptions of the orbitals. Note that in this point group (C_{2v}) the degenerate e' levels of the trigonal plane are split. With three pairs of electrons (BH_3, BY_3) there is no energy gain on distorting to the T shape. (For the three electron-pair case for which there is no structural data available a Jahn–Teller distortion is predicted. LiH_3^+, a

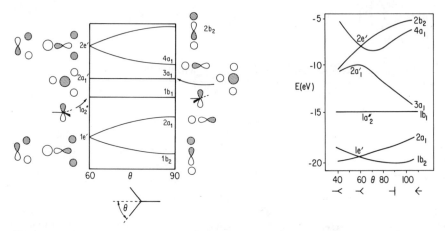

Figure 4.7   Energy changes of the orbitals of an AH_3 molecular on distorting the trigonal planar AH_3 species to a T shape (*a*) without and (*b*) with mixing between the orbitals of a_1 symmetry; (*b*) is constructed from the results of an EHMO calculation.

"mass spectrometer molecule," is calculated by *ab initio* methods to be Y-shaped.[175]) Neither is there any energy gain for the four (NY_3) and five (ClY_3) electron-pair cases. However when the a_1 orbitals are allowed to mix together, the situation changes. The mixing process is a little more difficult to appreciate here since there are three a_1 orbitals ($2a_1$, $3a_1$, and $4a_1$) to consider.

We are particularly interested in the behavior of the middle one of the trio, which is the HOMO in ClF_3. It is probably lowered more by mixing with the closer $4a_1$ orbital than elevated by mixing with $2a_1$, such that it is stabilized on distortion to the T shape. Figure 4.7*b* shows the result of an EHMO calculation. This argument concerning the merits of the two distortion routes in the three coordinate molecules involves the balancing act of an orbital mixing process. In view of this it is not surprising that several sets of numerical calculations have been performed on this sort of molecule in order to produce quantitative molecular orbital diagrams in which the orbital mixing process is automatically included.[67,73,74,95,168] This is especially true for molecules with higher coordination numbers where the number of angular degrees of freedom is larger and it is often impossible to choose between them using qualitative arguments. However it is important to be able to rationalize the features of such diagrams using qualitative orbital overlap and symmetry ideas as we have done here in order to understand the molecular orbital factors behind the actual choice of molecular geometry.

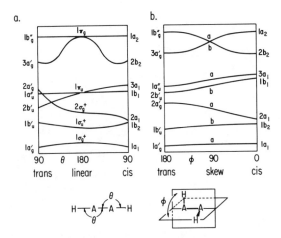

Figure 4.8 Energy changes of the orbitals of an A_2H_2 system on distortion. EHMO calculations on N_2H_2: (a) linear, cis, and trans planar geometries; (b) planar and nonplanar geometries. Adapted from Ref. 74.

Note both in Figure 4.2 and Figure 4.6 that the lone pair that is responsible for distorting the three electron-pair AH_2 system from linear and the four electron-pair AH_3 system from planar becomes stereochemically active automatically as the filled $2a_1$ orbital drops to lower energy. Thus the accomodation of the third and fourth electron pair respectively into the coodination shell around the A atom on the VSEPR scheme and the subsequent distorted geometry is matched in the molecular orbital approach by a stabilization of that orbital containing the lone pair on bending.

4.3 HAAH Systems

As an example of a more complex system we turn to HAAH molecules. This was a molecular type that Walsh did not analyze in simple terms. Even with just four atoms the correlation diagrams for bending the molecule are quite complex to be derived in a qualitative fashion. Figures 4.8a and 4.8b show the results of some EHMO calculations on the model species N_2H_2. We refer the reader to Ref. 74 for a dissection of the diagram and the reasons for the energy changes that are portrayed. With 10 electrons (C_2H_2) it is not immediately clear from the orbital diagram what geometry is preferred. Numerically however by weighting the calculated energy levels with the number of electrons in each orbital, the linear

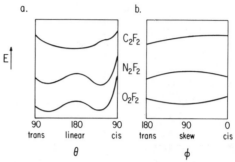

Figure 4.9 Variation in total orbital energy on distortion for a series of A_2H_2 molecules. Adapted from Ref. 73.

geometry wins out (Figure 4.9). With hindsight we could argue that on going to the trans planar structure the energy gain asssociated with $2b_u'$ is canceled out by the rise in energy of $2a_g'$ and so the energetics are controlled by $1b_u'$. In the cis planar configuration both $3a_1$ and $1b_2$ rise in energy on distortion. For the 12-electron system (N_2H_2) the stabilization of orbitals arising from π_g on bending makes the nonlinear geometry a possibility.

There is clearly little difference in energy between cis and trans planar geometries; if the nonplanar geometry were found it would almost certainly be a triplet. In fact N_2F_2 (N_2H_2 is not very stable) is a singlet, is planar and exists as cis and trans forms with a significant barrier between them (Figure 4.10). O_2H_2 (or O_2F_2) with 14 electrons again has a choice of nonlinear geometries open to it. The planar forms have one orbital derived from π_g stabilized on bending, the nonplanar form has two (Figure 4.8). It is not apparent from simple qualitative arguments which will be more stable but we recall from Section 2.3 that the geometry of the allene molecule was the one in which the ligands (CH_2) each made use of one of the central-atom π-type orbitals, so that the ligands avoid sharing the same central-atom orbital. The present case is an exactly analogous one. The observed nonplanar geometry may be understood on the basis that the interactions with two orbitals of Figure 4.11a are worth more than the one of Figure 4.11b, such that each ligand (H) makes use of different

Figure 4.10 Geometries of A_2Y_2 molecules.

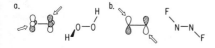

Figure 4.11 Two ligand σ orbitals (*a*) using different A_2 orbitals and (*b*) sharing the same A_2 orbital.

orbitals on the central unit (O_2). In fact the dihedral angle ϕ in the H_2O_2 molecule is sensitive to its environment. In $Na_2C_2O_4 \cdot H_2O_2$ it is 180°, so that the molecule is planar. A low-energy rearrangement process is therefore indicated here. Free O_2F_2 has a similar structure to H_2O_2 but with a dihedral angle of 87.5°. In these systems rather more quantitative ideas are necessary to be able to rationalize the geometries of these molecules, and this is a general feature of more complex polyatomic systems. Whereas the form of the correlation diagram—which orbitals are stabilized and which destabilized on distortion—is usually readily understandable, the energetic preferences for one arrangement over another need to be evaluated in a quantitative way.

5 Molecular Orbital Models of Main Group Stereochemistry: The Jahn–Teller Theorems

Symmetry arguments, as we have seen already, are of great use in looking at features of molecular structure. In this chapter we describe some symmetry rules for looking at molecular geometry which were initially formulated by Bartell and amplified by Pearson.[10,46,160,162] The model is based on the symmetry properties of the various terms arising in the perturbation expansion of the energy of a molecule on distortion along a normal coordinate of given symmetry species S_i. We shall inquire whether the energy is lowered if the symmetric structure is distorted and later on see how variation in ligand electronegativity affects the results. At the orbital level the second order Jahn–Teller approach turns out to be a simple perturbation viewpoint which can be used to comment on the slope of the HOMO. In this way it has clear ties with the Walsh approach of the previous chapter.

5.1 First- and Second-Order Jahn–Teller Effects[10,104,160–162]

We can simply write the energy of a molecule distorted within the coordinate S_i as a power series expansion of the form of Equation 5.1:

$$E(S_i) = E^0 + F_i S_i + \frac{1}{2} F_{ii} S_i^2 + \frac{1}{6} F_{iii} S_i^3 + \cdots \qquad (5.1)$$

E^0 is the energy of the undistorted geometry and the other parameters are the energy derivatives $F_i = (\partial E/\partial S_i)_0$, $F_{ii} = (\partial^2 E/\partial S_i^2)_0$, $F_{iii} = (\partial^3 E/$

$\partial S_i^3)_0$, and so on; F_{ii} is the vibrational force constant for distortion within this coordinate and F_{iii} is the cubic anharmonicity. The subscript 0 refers to the origin of the displacement coordinate S_i. Normally when we consider oscillations about equilibrium positions in molecules (molecular vibrations), we usually assume that the motion is harmonic and described by Hookes law $V = 1/2kS_i^2$, where k (identified with F_{ii} of Equation 5.1) is the vibrational force constant (positive) and S_i is the displacement from equilibrium. In this chapter however we often place molecules in geometries that are not the one of lowest energy and will need a more general description of the potential energy surface. The electronic Hamiltonian may be expanded in a similar fashion to the energy above (equation 5.2):

$$\mathcal{H}(S_i) = \mathcal{H}^0 + \mathcal{H}_i S_i + \frac{1}{2}\mathcal{H}_{ii}S_i^2 + \cdots \qquad (5.2)$$

where $\mathcal{H}_i$ and $\mathcal{H}_{ii}$ are analogously the derivatives $(\partial\mathcal{H}/\partial S_i)_0$ and $(\partial^2\mathcal{H}/\partial S_i^2)_0$, respectively. Using first-and second-order perturbation theory (Chapter 1) and collecting terms in S_i and S_i^2, we find that the energy expansion may be written as

$$E(S_i) = E_0 + S_i \int \Psi_0\mathcal{H}_i\Psi_0\,d\tau + \frac{1}{2}S_i^2 \left[\int \Psi_0\mathcal{H}_{ii}\Psi_0\,d\tau + 2\sum_m{}' \frac{\left| \int \Psi_0\mathcal{H}_i\Psi_m\,d\tau \right|^2}{\Delta E_{0m}} \right] + \cdots \qquad (5.3)$$

and the distorted wavefunction as

$$\Psi_0' = \Psi_0 + \sum_m{}' \frac{\int \Psi_0\mathcal{H}_i\Psi_m\,d\tau}{\Delta E_{0m}}\Psi_m S_i \qquad (5.4)$$

for the electronic ground state (Ψ_0). (The prime over the summation signs mean that $\Psi_m = \Psi_0$ is excluded from the summation over all the states Ψ_m of the undistorted molecule.) ΔE_{0m} (<0) is the zeroth-order energy separation between the states Ψ_0 and Ψ_m. The first-order function $S_i\int \Psi_0\mathcal{H}_i\Psi_0\,d\tau$ in Equation 5.3 represents the true or first-order Jahn–Teller term. If it is negative, then the geometry we are considering is not as stable as some other. It may be nonzero on group theoretical grounds if (a) Ψ_0 is nondegenerate (state of A or B symmetry) and S_i is of a_1 symmetry, (b) Ψ_0 is orbitally degenerate (e.g., a state of T symmetry of a tetrahedral geometry) and S_i is of a suitable symmetry to remove the degeneracy,* or of a_1 symmetry. The case of a_1 symmetry motions we ignore since these will not change the point group of the molecule. Case (b) then reduces to a statement of the (first-order) Jahn–Teller theorem— a nonlinear orbitally degenerate molecule will distort so as to remove this degeneracy.[104] For the most part we are not concerned with Jahn–Teller instabilities of this sort in main group molecules and therefore do not consider this term further here. We discuss the Jahn–Teller approach in greater depth in Chapter 11 when considering transition metal systems.

The force constant for distortion within this distortion coordinate, S_i, can be identified with the term in square brackets in Equation 5.3. It consists of two terms, a "classical" force constant $\int \Psi_0 (\partial^2 \mathcal{H}/\partial S_i^2)_0 \Psi_0 \, d\tau$ representing nuclear motion in a static electronic distribution and a relaxation term which describes the energy change associated with the electronic charge distribution as it adjusts to "follow" the nuclei (Figure 5.1). This force constant (Equation 5.5),

$$F_{ii}(0) = f_{00} + \sum_m{}' f_{0m} \quad (f_{0m} < 0) \tag{5.5}$$

may be positive or negative, depending upon whether the integral $\int \Psi_m \mathcal{H}_i \Psi_0 \, d\tau$ is nonvanishing (on group theoretical grounds) and whether the associated energy gap ΔE_{0m} is sufficiently small. If the force constant is negative, then the energy of the system must be lowered on distortion (Figure 5.2) and a second-order Jahn–Teller distortion away from the symmetrical geometry will be spontaneous. The molecule will distort until it finds a place with positive force constants for all nuclear displacements. On the other hand, the summation term may be nonzero but insufficiently large to overcome the positive "classical" force constant. In this case we expect to see a low or "softened" force constant associated with the distortion S_i.[5] The smallest energy gap ΔE_{0m} is clearly associated with the separation between the electronic ground state (Ψ_0) and the first excited electronic state of the same spin multiplicity (Ψ_1). In the present application of Equation 5.5, our attention is focused on the one term involving this state and hopefully the overriding term in the perturbation

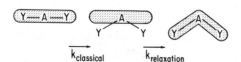

$k_{classical}$ $k_{relaxation}$

Figure 5.1 Classical and relaxation contributions to a vibrational force constant.

* Except for linear molecules where S_i can never be of π symmetry, the only symmetry which can lead to bending and hence removal of the degeneracy.

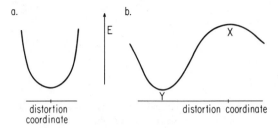

Figure 5.2 Schematic potential function for distortions of molecules: (*a*) positive force constant for distortion, (*b*) the negative force constant at *X* means the molecule will distort to *Y* where it finds a positive force constant for distortion.

expansion. This of course is nonrigorous in a mathematical sense but, as we shall see, is quite sufficient for our needs in viewing main group stereochemistry in this way. The integral $\int \Psi_0 \mathcal{H}_i \Psi_1 \, d\tau$ may be reduced from one involving many electron wavefunctions to one containing single electron wavefunctions by writing Ψ_0 and Ψ_1 as either single or combinations of Slater determinants. After a little algebraic manipulation it is found[5] that this integral contains the product $\psi_0 \cdot \psi_1$, where ψ_0 and ψ_1 are the molecular orbitals that hold the "excited" electron in the ground and first excited states. To determine when the effect may be important we need only to consider the symmetry species of this transition density $\psi_0 \psi_1$. The symmetry species of S_i needs to be identical to this for the integral $\int \Psi_0 \mathcal{H}_i \Psi_1 \, d\tau$ to be nonzero. If S_i corresponds to a normal bending mode of the molecule, then a second-order Jahn–Teller distortion is a possibility. From equation 5.4 we see that on such a distortion the HOMO is stabilized as HOMO and LUMO are mixed together. We therefore have a mechanism for determining which of the possible distortion modes of a molecule allows such mixing to occur, and this leads to an immediate correlation with the Walsh approach of the previous chapter. In essence what the second-order Jahn–Teller approach does is ask, for what symmetry distortion coordinate will the HOMO and LUMO mix together such that the HOMO is stabilized (and the LUMO destabilized) as in Figure

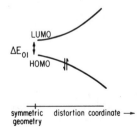

Figure 5.3 Alternative view of the second-order Jahn–Teller approach, which finds that geometric coordinate allows HOMO and LUMO to mix together.

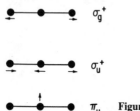

σ_g^+

σ_u^+

π_u **Figure 5.4** Normal vibrations of linear triatomic molecules. The π_u motion leads to a bent molecule.

5.3? The answer may be, no distortion at all (as we shall see for BX_3) or one of the angular degrees of freedom open to the molecule (as we shall see for NH_3 and ClF_3). The group theoretical analysis is a method finding the symmetry species of this coordinate.

5.2 AY_2 Molecules

The normal modes of vibration of these linear triatomic species are shown in Figure 5.4. We have derived a molecular orbital diagram for the linear molecule before (Figure 4.1). The application of the second-order Jahn–Teller method to these systems is shown in Figure 5.5, for BeY_2, OY_2 and XeY_2 species. For BeY_2 the transition density is of species $\sigma_u^+ \times \pi_u = \pi_g$. No normal vibrational mode exists with this symmetry. Since the energy gaps ΔE_{0m} are all large and since the smallest gap contributes nothing to the lowering of force constant, BeY_2 must have a positive force constant for bending away from the linear geometry, so that it should be linear.

For CY_2 and OY_2 the situation is very different. Here the transition density is of species $\pi_u \times \sigma_g^+ = \pi_u$, which does correspond to the bending

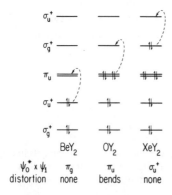

Figure 5.5 Application of the second-order Jahn–Teller method to test for the sign of the bending force constant at the linear geometry for AY_2 molecules.

coordinate for AY_2. Since the energy gap here may well be small, the force constant for bending the linear geometry is expected to be negative, and these molecules should be bent. For XeY_2 this $\pi_u \rightarrow \sigma_g^+$ "transition" is now blocked, but the $\sigma_g^+ \rightarrow \sigma_u^+$ "transition" needs to be considered. This gives a transition density of species σ_u^+ which corresponds to an AY_2 vibrational coordinate but not a bending one (Figure 5.4). A linear molecule is thus expected but with a reduced frequency (force constant) representing antisymmetric stretching (ν_3). As a rule, since bending force constants are much lower than stretching force constants, the second-order Jahn–Teller interactions will tend to soften or lower stretching force constants but lead to negative bending force constants. Following standard practice in molecular vibration theory we write two functions which relate the symmetric (ν_1) and antisymmetric (ν_3) stretching frequencies of a linear triatomic molecule (Equations 5.6):

$$\nu_{asym}^2 = aF_{33} = a(f_r - f_{rr})$$

$$\nu_{sym}^2 = bF_{11} = b(f_r + f_{rr})$$

(5.6)

The terms a and b are readily determined functions of the atomic masses, f_r is the AY stretching force constant, and f_{rr} is the bond–bond interaction force constant. We may also express the force constants after the style of Equation 5.5 by including a relaxation contribution from the first excited state. Since there is a relaxation contribution from this source for $F_{33}(\sigma_u^+)$ but not for $F_{11}(\sigma_g^+)$, we may write

$$\nu_{asym}^2 = a(f_{00} + f_{01})$$

$$\nu_{sym}^2 = b(f_{00})$$

(5.7)

A little rearrangement gives two Equations 5.8 of the same form as Equations 5.6:

$$\nu_{asym}^2 = a\left(\frac{2f_{00} + f_{01}}{2} + \frac{f_{01}}{2}\right)$$

$$\nu_{sym}^2 = b\left(\frac{2f_{00} + f_{01}}{2} - \frac{f_{01}}{2}\right)$$

(5.8)

Since f_{01} is negative this implies that the bond–bond interaction force constant is positive, which is indeed the case for XeF_2. KrF_2, which should be similar, has a negative f_{rr}. It has been suggested that the presence of very weak and long Kr–F bonds invalidate the use of simple single configuration molecular orbital ideas here. In general however the use of the

Table 5.1 Some Bond–Bond Interaction Constants in Triatomic and Tetratomic Systems

Molecule	Symmetry of Transition Density	Predicted Sign of f_{rr}	Observed $f_{rr}(\mathrm{Nm}^{-1})^a$
CO_2	Σ_u^+	+	130
CS_2	Σ_u^+	+	60
N_3^-	Σ_u^+	+	160
N_2O	Σ^+	+	136
COS	Σ^+	+	200
ClCN	Σ^+	+	100
UO_2^{2-}	Σ_g^+	−	−30
NpO_2^{2-}	Σ_g^+	−	−54
PuO_2^{2-}	Σ_g^+	−	−89
AmO_2^{2-}	Σ_g^+	−	−166
NO_2	B_2	+	145
SO_2	B_2	+	2.4
O_3	B_2	+	152
ClO_2	B_2	+	±
OCl_2	B_2	+	+
OF_2	B_2	+	110
H_2O	A_1	−	10
H_2S	A_1	−	−22
H_2Se	A_1	−	−25
$BO_2(\tilde{X})$	Σ_u^+	+	142
$BO_2(\tilde{A})$	Σ_u^+	−	−89.5
$CO_2^+(\tilde{X})$	Σ_u^+	+	196
$CO_2^+(\tilde{A})$	Σ_u^+	−	−156
BF_3	E'	+	78
BCl_3	E'	+	31
BBr_3	E'	+	61
PF_3	E	+	39
PCl_3	E	+	27
PBr_3	E	+	3
AsF_3	E	+	34
$SbCl_3$	E	+	16
$BiCl_3$	E	+	15
$AsCl_3$	E	+	19

^{a} Adapted from Ref. 5

second-order Jahn–Teller approach to view the signs of these interaction constants is very successful (Table 5.1).

The opposite behavior of HOMO and LUMO on distortion suggests that the properties of the ground and first excited states ought to be quite different. This effect is most easily seen expressed as Equation 5.9,

$$F_{ii}(1) = f_{11} + f_{10} = f_{11} - f_{01} \qquad (5.9)$$

to be compared with the analogous equation for the ground state (Equation 5.10):

$$F_{ii}(0) = f_{00} + f_{01} \qquad (5.10)$$

For the excited state the relaxation term of Equation 5.8 is the same size as for the ground state but of the opposite sign ($\Delta E_{10} = -\Delta E_{01}$). Thus a sign reversal in the bond–bond interaction constant between ground and excited states is expected, which is indeed found for BO_2 and CO_2^+ in Table 5.1, two systems where data are available. This is assuming of course that the first excited state is not strongly perturbed by another (higher-energy state) on distortion.

The influence of ligand electronegativity on the geometry arises naturally from this approach. Increasing the electronegativity of Y relative to A lowers the energy of the $Y\sigma$ atomic orbitals relative to A. This has the effect of lowering the entire set of σ orbitals relative to $1\pi_u$ (σ nonbonding at the linear AY_2 geometry). The result is a decrease in the energy seperation ΔE_{01} between $1\pi_u$ and $2\sigma_g^+$ orbitals and makes bending of the molecule even more advantageous. For the OY_2 series we now have a way to view the bond angle trends in the series OF_2 (103°), OH_2 (104.5°), $O(SiH_3)_2$ (144°), and OLi_2 (180°). Inclusion of π-type orbitals on the ligands also adjusts this $1\pi_u$–$2\sigma_g^+$ energy gap. If the ligand orbitals lie to lower energy than the central atom orbitals (e.g., OF_2), then this energy separation is reduced; if they lie to higher energy (e.g., OLi_2), then this gap is increased and bending is less advantageous (Figure 5.6).

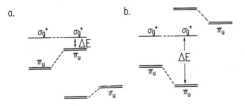

Figure 5.6 Influence of ligand π orbitals on the size of the second-order Jahn–Teller effect by changing the HOMO–LUMO gap (ΔE): (*a*) π donors; (*b*) π acceptors.

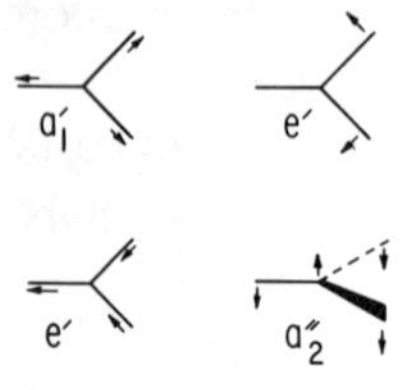

a_1' e'

e' a_2''

Figure 5.7 Normal vibrations of a trigonal planar (D_{3h}) AY$_3$ molecule. The a_2'' and e' bending modes lead to trigonal pyramidal and T-shape geometries, respectively.

5.3 AY$_3$ Molecules

The analysis of these molecules follows along similar lines to the AY$_2$ system above. We start with a symmetrical D_{3h} geometry. Figure 5.7 shows the normal modes of such a species—two of these coordinates change the angular geometry of the molecule. Figure 5.8 shows the operation of the scheme for molecules having three to five electron pairs. For BY$_3$ the transition density corresponding to the lowest energy transition is of species $e' \times a_2'' = e''$, which does not describe a normal mode and the molecule remains undistorted. For NY$_3$ a distortion coordinate of species a_2'' is found which sends the molecule to the observed C_{3v} pyramidal geometry. For ClF$_3$ the distortion mode is of species e' and the T shape results. If the e' stretching and bending coordinates take part in the motion, we can see how the two symmetry-unrelated bond lengths in the T-shape molecule become different in length. The same electronic effect that made the force constant F_{44} negative will contribute to make the interaction force constant F_{34} negative too. This means that the e' species stretching and bending coordinates must have the same sign for the potential energy to be lowered on distortion. Thus opening up one angle of the trigonal plane (e' bend) will result in a lengthening of the two bonds defining that angle (e' stretch). This is of course just the VSEPR conclusion; bonds adjacent to lone pairs are longer than bonds which are not.

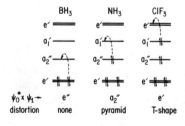

Figure 5.8 Application of the second-order Jahn–Teller method to test for structural instability of trigonal planar AY$_3$ molecules.

5.4   AY₄ and AY₅ Molecules

Figures 5.9 and 5.10 show a similar approach for the AY_4 molecules. The four electron-pair tetrahedral CY_4 molecule is stable to second-order Jahn–Teller distortion not because the transition density is of the wrong symmetry but because the $1t_2-2a_1$ energy gap is large. The five electron-pair SF_4 molecule is unstable at this geometry because the energy gap $2a_1-2t_2$ is much smaller and this species distorts via a t_2 distortion to its characteristic butterfly shape (Figure 5.9). XeF_4, with six pairs of electrons, is first-order Jahn–Teller unstable (two electrons in $2t_2$) and distorts via an e species vibration to a square plane.

For five-coordinate molecules we are interested in the relative stabilities of the square pyramid (spy) and trigonal bipyramid (tbp). If the molecular orbital ordering of Figure 5.11a is accepted then tbp PF_5, with five pairs, is stable to bending (transition density of species a_1'), but BrF_5, with six pairs, is unstable to bending with a transition density of species e' (Figure 5.11b). Different results will hold for alternative orbital ordering. If $3a_1'$ lies above $2e'$ and $2a_2''$, then PF_5 is second-order Jahn–Teller unstable and (singlet) BrF_5 is unstable at the tbp geometry on first-order Jahn–Teller grounds. In order to understand and rationalize the geometries of molecules with these coordination numbers, we need to know the correct orbital order of the symmetrical structure and have some qualitative ideas as to energy separations.

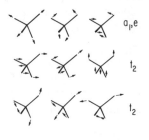

Figure 5.9 Normal bending and stretching vibrations of a tetrahedral AY_4 molecule.

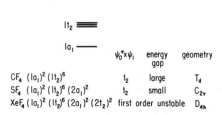

	$\psi_0^* \times \psi_1$	energy gap	geometry
CF_4 $(1a_1)^2 (1t_2)^6$	t_2	large	T_d
SF_4 $(1a_1)^2 (1t_2)^6 (2a_1)^2$	t_2	small	C_{2v}
XeF_4 $(1a_1)^2 (1t_2)^6 (2a_1)^2 (2t_2)^2$	first order	unstable	D_{4h}

Figure 5.10 Molecular orbital diagram for a tetrahedral AH_4 molecule and the Jahn–Teller rationalizations of the geometries of CF_4, SF_4, and XeF_4.

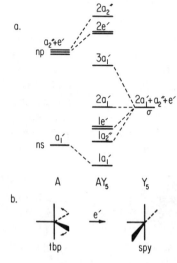

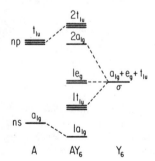

Figure 5.11 (a) Molecular orbital diagram for a trigonal bipyramidal (D_{3h}) AH_5 molecule (or the σ framework of an AY_5 molecule). (b) One component of the e' species bending vibration of a trigonal bipyramidal molecule showing how the square pyramidal structure is reached.

5.5 AY_6 Molecules

A molecular orbital diagram for an AY_6 molecule is shown in Figure 5.12. The normal modes transform as $a_{1g} + e_g + t_{1u}$ (stretching) and $t_{1u} + t_{2g} + t_{2u}$ (bending). Application of the second-order Jahn–Teller ideas leads to a prediction of the observed octahedral geometry for SF_6 (six pairs) with a transition density of species e_g but a distorted geometry for XeF_6 (seven pairs). Here a transition density of species t_{1u} is found ($2a_{1g} \rightarrow 2t_{1u}$). A t_{1u} bending motion behaves as if a lone pair were pushing its way out of one face of the octahedron, and indeed this molecule experimentally has been shown to possess a low-frequency vibration which allows the lone pair to "visit" each of the eight symmetry-related posi-

Figure 5.12 Molecular orbital diagram for an octahedral (O_h) AH_6 molecule (or the σ framework of an AY_6 molecule).

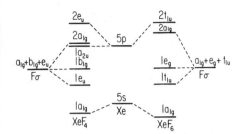

Figure 5.13 σ-Only molecular orbital diagrams of square planar XeF_4 and octahedral XeF_6. Note the smaller $2t_{1u}$–$2a_{1g}$ energy gap in XeF_6 compared to the $2e_u$–$2a_{1g}$ gap in XeF_4.

tions of the capped octahedral structure (Figure 3.5). The molecule is certainly nonoctahedral in the gas phase, but the size of the distortion observed is much smaller than that expected on the VSEPR scheme and smaller than the gross distortions from the symmetrical structure found for ClF_3, for example.

Even more interesting is a comparison between XeF_4 and XeF_6. Before, we examined the Jahn–Teller susceptibility of XeF_4 at the tetrahedral geometry. Here we look at the square planar (i.e., experimental) geometry. The pattern of the molecular orbital diagram is very similar to that for octahedral XeF_6. In the XeF_6 case there is a second-order Jahn–Teller transition from $2a_{1g}$ to $2t_{1u}$; for XeF_4 there is an analogous transition from $2a_{1g}$ to $2e_u$ (Figure 5.13), giving rise to a transition density of species e_u. A motion of this symmetry results in the protrusion of a lone pair in plane between two fluorine atoms (Figure 5.14). XeF_4 is however square planar (with perhaps a low frequency e_u bending mode), and the difference in behavior between the two molecules (one flexible, the other rigid) is ascribed to the larger energy gap in the four compared to the six-coordinate molecule. Relative to the atomic $5p$ orbital the $2e_u$ orbital (in XeF_4) and the $2t_{1u}$ orbital (in XeF_6) are at the same energy. They both arise via the out-of-phase overlap of two trans-fluorine σ orbitals with a central-atom p orbital. The e_u orbital contains two such interactions (degenerate), and the t_{1u} orbital contains three (Figure 5.12). However the $2a_{1g}$ orbital is an out-of-phase combination of the central-atom s orbital with the ligand σ orbitals and will be pushed to higher energy in XeF_6 than in XeF_4. This may readily be seen by employing the angular overlap model introduced in Section 2.5. The overlap integral between ligand a_{1g} combinations and the central-atom s orbital is from

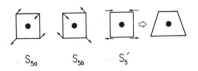

S_{5a} S_{5b} S_5'

Figure 5.14 E_u distortion of square planar XeF_4. A linear combination of S_{5a} and S_{5b} (S_5') leads to a distortion where a lone pair might be envisaged as pointing between two ligands.

group theoretical considerations $\sqrt{6}S_\sigma$ and $\sqrt{4}S_\sigma$, respectively, for octahedral and square planar geometries. This leads to a larger a_{1g} interaction in the octahedral ($6e_\sigma$) compared to square planar ($4e_\sigma$) case.

The $2a_{1g}$–$2e_u$ gap should then on this simple model be larger than the $2a_{1g}$–$2t_{1u}$ gap. As noted above, use of second-order Jahn–Teller or perturbation arguments in general in these larger molecules does need a feeling for the sizes of the relevant energy gaps. Alternatively given a molecular geometry we may comment on the size of relevant energy seperations of the molecular orbital diagram.

5.6   The Case of CH₃

We recall that the methyl and substituted methyl radicals occupied a rather special place in Walsh's scheme for AH_3 molecules. Whereas it was conceivable that two electrons in the $2a_1$ orbital of Figure 3.6 would produce a pyramidal molecule (NH_3), it was not clear whether one electron was sufficient to give rise to the same effect. On the VSEPR scheme the same question was couched in a different way—how does half an electron pair behave? The experimental data show planar CH_3 but pyramidal substituted radicals. Use of the second-order Jahn–Teller approach shows that the integral $\int \Psi_0 \mathcal{H}_1 \Psi_1 \, d\tau$ may be nonzero for the a_2'' out-of-plane bending mode for both CH_3 and CY_3 molecules. The situation is quite analogous to NH_3, but it may be shown that because of the difference in the number of electrons in $1a_2''$ the relaxation part of the force constant will be half as large as for NH_3. Thus either a lowered out-of-plane bending force constant F_{22} and a planar molecule or a pyramidal species is expected. For CH_3 the former is found, but as the ligand electronegativity increases the energy gap ΔE_{01} decreases (as noted for the analogous AY_2 case) and the tendency to become pyramidal increases. Isoelectronic BrO_3 is also pyramidal.[15]

There are in addition to these observations two further interesting features concerning these radicals.[29]

1.   The out-of-plane bending vibration v_2 in the free CH_3 radical is very anharmonic (Table 5.2) such that the ratio v_H/v_D is quite different from its usual value. Additionally the anharmonicity is in the opposite direction to that usually found in such molecules. The quartic term F_{2222} in the energy expansion of Equation 5.1 (the cubic term F_{222} is zero by symmetry) is positive (negative anharmonicity) rather than negative (positive anharmonicity) as usually found.

Table 5.2 Vibrational Data for CH_3 and Perturbed CH_3 Radicals

	ν_2 (cm^{-1})	ν(H)/ν(D)^a	
CH_3 . . . LiBr	730	1.288	Normal positive anharmonicity
CH_3 . . . LiI	730		
CH_3 . . . NaBr	700		
CH_3 . . . NaI	696	1.305	
CH_3 . . . KI	680		Negative anharmonicity
CH_3	611	1.319	

^{a} 1.291 for harmonic oscillator.

2. The CH_3 radical may be made in a low-temperature matrix perturbed by an alkali halide molecule. Table 5.2 shows that in the presence of the AX molecule the value of ν_2 (and hence the force constant) is shifted upward with an associated decrease in the negative anharmonicity and the eventual substitution of a small (normal) positive anharmonicity.

The effect of alkali halide perturbation is probably to increase the energy gap ΔE_{01} between $1a_2''$ and $2a_1'$ orbitals (Figure 5.15) by σ interaction between the carbon p_z orbital and a directional orbital on the AX unit. Thus on increasing interaction the second-order Jahn–Teller softening term decreases and the vibrational frequency increases. Just as this force constant may be written as the difference of two terms (Equations 5.3 and 5.5), we can show that the quartic anharmonicity constant consists of classical (negative) and relaxation (positive) contributions. The latter is simply written as Equation 5.11 for the CH_3 radical if only one excited state is included:

$$- \frac{\int \Psi_0 \mathcal{H}_i \Psi_1 \, d\tau}{\Delta E_{01}} \cdot \frac{\int \Psi_1 \mathcal{H}_{ii} \Psi_1 \, d\tau}{(\Delta E_{01})^2} \tag{5.11}$$

Thus as the vibrational softening of the out-of-plane bending mode increases ($\int \Psi_0 \mathcal{H}_2 \Psi_1 \, d\tau/\Delta E_{01}$ increases), F_{22} drops to lower frequency and F_{2222} becomes positive, as found experimentally in Table 5.2.

By the use of these quite simple ideas we can describe and understand some rather subtle features of the potential energy surface of a molecule.

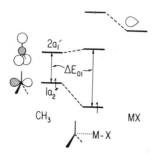

Figure 5.15 Perturbation of CH_3 energy levels by a close alkali halide molecule, showing the increase in $1a_2''-2a_1'$ energy gap.

Interesting effects of a similar nature ought also to be observed in other "halfway" molecules such as the unknown OF_3 at the D_{3h} geometry with one electron in the $3a_1$ orbital. A similar negative anharmonicity in the free molecule is observed experimentally in the first excited electronic state of NH_3 $((1e')^4(1a_2'')^1(2a_1')^1$, which is also planar. The explanation is exactly analogous to the one here. BH_2 occupies a similar position to CH_3 in the triatomic system. However the molecule is nonlinear. Our arguments for CH_3 only apply to the ground electronic state of course. However from the perturbation theory arguments, if CH_3 is only tenuously planar in the ground state it ought to be much more strongly planar in its first excited state. The out-of-plane bending frequencies (v_2) accordingly are respectively 580 and 1360 cm^{-1} for the ground and excited states from the gas-phase electronic spectrum.

5.7   HAAH Systems

The key feature of Walsh's scheme was that the lower-lying molecular orbitals either changed slightly in energy or were destabilized on distorting a symmetrical geometry. Eventually on moving to higher energies one comes to a pair of orbitals of different symmetry in the symmetrical reference geometry but which strongly mix together as the molecule distorts such that the lower-energy one is rapidly stabilized and the higher-energy one is rapidly destabilized. Occupation of the lower orbital of this pair resulted in a stabilizing contribution on distortion which, if large enough, could overcome any destabilization contributed by occupied lower orbitals. The magnitude of the stabilization using the second-order Jahn–Teller approach depends upon the number of electrons in the HOMO just as in a straightforward energy summation procedure. We noted that the effect was twice as large for ammonia as for CH_3, for example. The general success of the scheme suggests that it is perhaps the energetic behavior

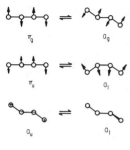

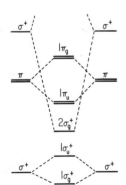

Figure 5.16 Bending vibrations of A_2H_2 molecules in linear, cis, and trans planar and skew geometries.

of the HOMO on distortion that controls the geometry. However we now look at H_2A_2 molecules where, as described in Section 4.3, the geometry is not clearly dictated by the energetics of the HOMO. For these species the second-order Jahn–Teller scheme is not very good.

We note linear C_2H_2 (C_2F_2), two different bent but planar geometries for N_2F_2 (cis and trans) and a nonplanar bent geometry for O_2H_2 (O_2F_2) (Figure 4.10). Figure 5.16 shows the normal modes of vibration for various A_2H_2 systems. For the linear molecule an orbital level ordering is shown in Figure 5.17, which we may readily construct from the orbital pictures of two AH molecules (see the diagram for HF in Figure 2.7). For acetylene with the configuration $(2\sigma_g^+)^2(2\sigma_u^+)^2(3\sigma_g^+)^2(1\pi_u)^4$ the species of the transition density is $\pi_u \times \pi_g = \delta_u + \sigma_u^+ + \sigma_u^-$, which does not contain any of the bending modes of Figure 5.16. N_2H_2 (N_2F_2) with a low-lying $\pi_g \rightarrow \sigma_u^+$ transition and a transition density of species π_u is predicted to have a planar cis bent configuration (N_2F_2 is found in both cis and trans structures).

Figure 5.17 Molecular orbital diagram of linear A_2H_2 molecule assembled from two AH fragments.

An analogous argument leads to the same prediction for O_2H_2. This latter result is not a failure of the approach since there is not a single bending mode that will give the nonplanar bent geometry. What we need to do now is test this cis bent geometry itself for second-order Jahn–Teller stability. From Figure 4.8 the lowest-energy transition leads to a transition density of species $a_2 \times b_1 = b_2$. This is not the species of vibration needed to send the O_2H_2 geometry to the nonplanar skew geometry (a_2). The approach in this case is clearly not very satisfactory and is certainly connected with the fact that the energy changes associated with the HOMO alone may not always govern the choice of molecular geometry as we discussed in the previous chapter. However we are able to comment on the substantial energy barrier between cis and trans N_2F_2. In the cis geometry an a_2 vibration is needed to rotate one half of the molecule against the other (Figure 5.16). The transition density arising from the lowest energy transitions are b_1 and a_1. For the trans geometry an a_u vibration is required. The corresponding transition densities are a_g and b_u. Thus no softening of the force constant associated with the isomer interconversion is predicted, and the barrier may be large. (There are of course other ways of interconverting cis and trans isomers which require combinations of symmetry modes we cannot treat in this way.)

One of the excited low-lying electronic states of the acetylene molecule arises through promotion of an electron (real, not imaginary, as on the second-order Jahn–Teller scheme) from π_g to π_u. The molecule has the planar trans configuration—the cis would be predicted on the present approach.

5.8 Secondary Coordination in Nonmetallic Elements

In many main group systems the crystalline molecular structure shows that a molecule, defined by several short internuclear distances (bonds) around a central atom, often contains several other central atom-ligand distances that are much longer than normal bonded contacts but shorter than the sum of the relevant van der Waals radii. Sometimes in a series of related structures the change from an intramolecular to intermolecular contact may be viewed overall as a gradual one. Figure 5.18a shows the variation[32] in the two I–I distances in an I_3^- unit. At $r_1 = r_2 = 2.90$ Å the molecule is clearly best described as I_3^-, but when $r_1 = 4.0$ Å and $r_2 = 2.67$ Å the system is best described as an I_2 molecule perturbed by an I^- ion. These molecules may be classified very simply in terms of the total number of electron pairs (N) around the central atom in the extended structure, the number of short (S) and long bonds (L), and the number of lone

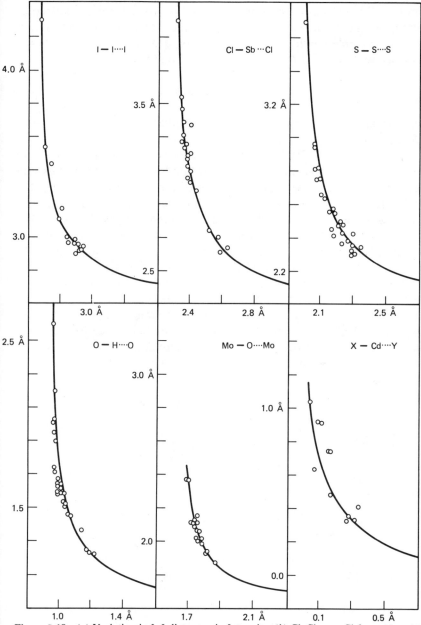

Figure 5.18 (*a*) Variation in I–I distances in I_3^- units; (*b*) Cl–Sb . . . Cl fragments; (*c*) 6*a* thiathiophthenes; (*d*) hydrogen bridges O–H . . . O; (*e*) Mo–O . . . Mo in molybdenum oxides; (*f*) X–Cd . . . Y distances. Reprinted with permission from H.-B. Bürgi, *Angew. Chem.* **14** 462 (1974).

pairs (P). P is of course the same whether we consider the "molecule" as comprising only the short bonds or if all are included. The central atom in the I_3^- example would be described by $N = 5$, $S = 1$, $L = 1$, $P = 3$, which we write as $NSLP = 5113$. Clearly Equation 5.12 holds,

$$N = S + L + P \tag{5.12}$$

and there are several ways of making up a given N with different values of S, L, and P.

The structures of ClF_2^+ and BrF_2^+ in $XF_2^+SbF_6^-$ contain (Figure 5.19) two short internuclear contacts (normal bonds) and two long ones with internuclear distances less than the sum of the van der Waals radii (for F, Cl, and Br the radii are 1.47, 1.75, and 1.85 Å). XF_2^+, with a total of four electron pairs, is expected to have a C_{2v} geometry, as observed. With two extra ligands (two extra pairs) there are now a total of six pairs, and a square planar structure should result as indeed is the case. Here $NSLP = 6222$. In $SbCl_3$, · $C_6H_5NH_2$ ($NSLP = 5311$) there is a contact shorter than the sum of van der Waals radii between the N and Sb atoms such that the overall geometry (with $N = 5$) resembles that of SF_4. The equivalence of the three bond lengths expected for an isolated C_{3v} $SbCl_3$ unit is removed, the "axial" one of the C_{2v} structure being lengthened. Some more examples are shown in Table 5.3. Figure 5.20 and Table 5.4 show how the structures may be envisaged, using the VSEPR approach. With

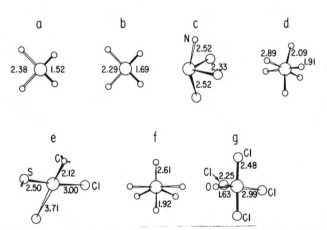

Figure 5.19 Some examples of molecules containing both short and long central-atom–ligand contacts: (*a*) ClF_2^+ in $ClF_2^+SbF_6^-$; (*b*) BrF_2^+ in $BrF_2^+SbF_6^-$; (*c*) $SbCl_3$ in $SbCl_3 \cdot C_6H_5NH_2$; (*d*) tetragonal TeO_2; (*e*) $C_6H_5TeSC(NH_2)Cl$; (*f*) SbF_3; (*g*) $SeOCl_3^-$ in $(C_5H_6N^+)_2SeOCl_3^-Cl^-$. Bond lengths in Å.

	Linear Group Y–A...Y'	Distance (Å)		Normal (Å)	van der Waals[b]
		YA	A...Y'	A–Y	A...Y'
$XeF^+Sb_2F_{11}^-$	F Xe F	1.84	2.35	1.93	3.47
$K^+XeO_3F^-$	O Xe F	1.76	2.48	1.93	3.47
$ClF_2^+SbF_6^-$	F Cl F	1.57	2.33	1.63	3.22
$BrF_2^+SbF_6^-$	F Br F	1.69	2.29	1.75	3.32
$Li^+IO_3^-$	O I O	1.81	2.89	1.94	3.50
I_2	I I I	2.72	3.50	2.56	3.96
$(C_6H_5)_2I^+Cl^-$	C I Cl	2.08	3.08, 3.24	2.27	3.73
TeO_2	O Te O	1.91	2.89	1.98	3.58
$SeOCl_3^-$	O Se Cl	1.59	3.38	2.13	3.65
$C_9H_8NO^+$	Cl Se Cl	2.27, 2.23	2.96, 2.99	2.13	3.65
$SeF_3^+NbF_6^-$	F Se F	1.69, 1.73	2.33, 2.43	1.78	3.37
TeF_4	F Te F	1.87	2.26	1.96	3.53
$(CH_3)_3Te^+$					4.04
$CH_3TeI_4^-$	C Te I	2.01	3.97	2.60	4.04
SbF_3	F Sb F	1.90	2.63	2.00	3.52
$AsF_3 \cdot SbF_5$	F As F	1.64	2.73	1.82	3.32
$SbCl_3 \cdot C_6H_5NH_2$	Cl Sb N	2.52	2.53	2.06	3.60
$(CH_3)_3GeCN$	C Ge N	1.98	3.57	1.92	3.50
$ClCN$	Cl C N	1.57	3.01	1.69	3.30
$XeF_5^+PtF_6^-$	F Xe F	1.89	2.52, 2.95	1.93	3.47

[a] Adapted from N. W. Alcock, *Adv. Inorg. Radiochem.* **15** 1 (1972).

[b] A reassessment of van der Waals radii is presently underway as a result of structural information on van der Waals molecules in supersonic jets. The Ar–Cl distance in Ar . . . ClF for example is very close to that found for the Cl . . . Cl distance in solid chlorine.

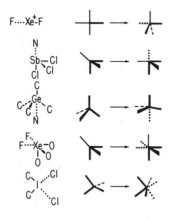

Figure 5.20 Schematic coordination geometries of some of the molecules of Tables 4.3 and 4.4 viewed using VSEPR. The heavy lines indicate electron pairs between primary bonded atoms; dotted lines, electron pairs for secondary coordination; and light lines, lone pairs.

a total of seven pairs the geometries still seem to be basically octahedral, but it is difficult to see how the angular geometry is distorted when the structure is already distorted by having one or more long bonds. One example of a molecule caught in the act is the story of trans $TeCl_4(thiourea)_2$. The orthorhombic form was found to have an octahedrally based centro-symmetric geometry (Figure 5.21). Over a period of a few years the crystals had changed to a monoclinic form and the structure of the molecule had a significantly different geometry, both in terms of bond lengths and bond angles.[57] This new structure has $NSLP$ = 7421 and, if the two extra ligands were completely dissociated, the SF_4 geometry would result. Interestingly the bond angle between the long bonds has increased (orthorhombic → monoclinic) a distortion in the opposite direction expected

Table 5.4   Some of the Molecules
Illustrated in Figures 5.19 and 5.20

	NSLP
$XeF^+[Sb_2F_{11}^-]^a$	5113
$SbCl_3NH_2C_6H_5$	5311
$(CH_3)_3GeCN$	5410
$(C_6H_5ICl)_2$	6222
$(XeO_3F^-)_n[K^+]_n$	6321
SbF_3	7331
TeO_2	7421
$CH_3TeI_4^-[(CH_3)_3Te^+]$	6511
$XeF_5^+[PtF_6^-]$	(10)541

^{a} Counterion if present in brackets.

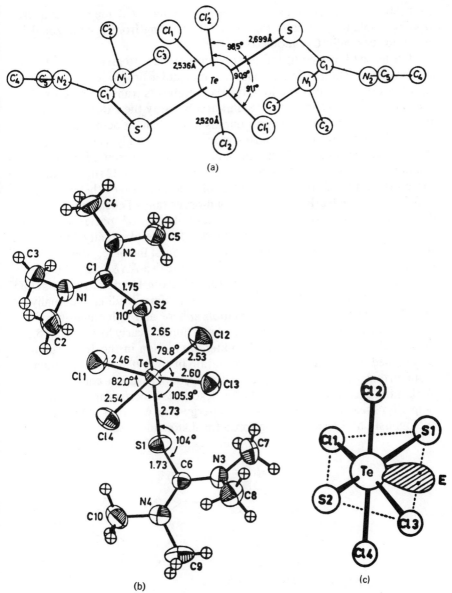

Figure 5.21 Structural changes in substituted trans $TeCl_4(thiourea)_2$ showing the tellurium coordination in (*a*) orthorhombic (reprinted with permission from S. Husebye and J. W. George, Inorg. Chem. **8** 314 (1969). Copyright by the American Chemical Society) and (*b*) monoclinic modifications; (*c*) Shows the implied position of the protruding lone pair (reprinted with permission from S. Esperås, J. W. George, S. Husebye, and Ø. Mikalsen, *Acta Chem. Scand.* **29** 141 (1975)).

from steric considerations. This has been cited as evidence for a protruding in-plane lone pair (Figure 5.21) resulting from a pentagonal bipyramid, basically C_{2v} geometry.

Let us see how the ligand dissociation process to give long and short internuclear distances is treated using molecular orbital ideas. Note that for all these structures the long bonds are always trans to some or all of the short bonds in the molecule. Put another way the extended molecule (long and short bonds) always contains more than an octet (four valence pairs) of electrons. These species may then be readily viewed using the three-center approach of Rundle and Pimentel augmented by s-orbital interaction as we describe in the next chapter. Figure 5.22 shows a familiar three-orbital picture for the p orbitals of this species (see the allyl radical in Section 2.3). Clearly there is a second-order Jahn–Teller transition from $1\sigma_g^+$ to $2\sigma_u^+$ which gives rise to a transition density of species σ_u^+ leading to a soft antisymmetric stretching mode. This is of exactly the correct form for contraction of one bond and extension of the other, the distortion observed experimentally both in I_3^- and the iso-$NSLP$ case of XeF^+ (Figure 5.20). Thus all three-center four-electron bonds are in principle susceptible to asymmetric distortions of this type, leading eventually to ligand expulsion from the coordination sphere and tightening up of the trans bond. The driving force for the distortion is clearly loss (Figure 5.22) of central-atom s–ligand σ antibonding interaction in the $1\sigma_g^+$ orbital to be weighed against the loss of stabilization in $1\sigma_u^+$. If s-orbital interaction is important then the driving force for asymmetrization could be large. Within the second-order Jahn–Teller description increasing s-orbital participation means that the $1\sigma_g^+$–$2\sigma_u^+$ energy gap will be reduced leading analogously to a larger driving force for distortion.

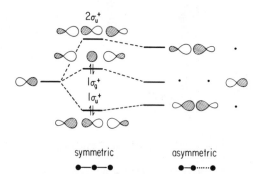

Figure 5.22 Molecular orbital diagram for symmetrical and asymmetric A_3 species. While the $1\sigma_g$ orbital will decrease in energy on asymmetrization, the behavior of $1\sigma_u$ will be system dependent. Here we show the case where it is destabilized during this process.

Figure 5.23   One component of the t_{1u} stretching mode of the octahedral AY_6 molecule showing the shortening of one bond and the lengthening of the bond trans to it.

For the seven-pair case we saw earlier that at the octahedral geometry a second-order Jahn–Teller distortion of·species t_{1u} was predicted. The size of the distortion was suggested to be controlled to a large extent by the importance of As–ligand σ interaction which raised the $2a_{1g}$ orbital in energy and decreased the $2a_{1g}$–$2t_{1u}$ energy gap. For this molecule there is however also a triply degenerate stretching vibration of t_{1u} symmetry (Figure 5.23), the analog of the σ_u^+ vibration of I_3^-. This suggests that the stereochemical activity of the seventh pair and the presence of long bonds are intimately connected. In the $TeCl_4$ tu$_2$ example (Figure 5.21) we see a rare situation where their appearance together can be observed experimentally. If s-orbital participation is the key factor then we would expect the importance of the effect to be greatest for the heavy elements. Te rather than Se, I rather than Br or Cl, Bi and Sb rather than As and Xe, and with electronegative ligands, for example, O, F with high p-orbital ionization potentials. There is in fact a general prevalence of this sort of complex in the examples described in Figures 5.19 and 5.20. This type of behavior has been described over the years as the "inert pair" effect,[52] as an explanation or description of the reluctance of the ns^2 pair of electrons in some of these heavy elements to exert a stereochemical influence in the molecule. In general undistorted (octahedral) seven-pair molecules have significantly longer bond lengths than their six-pair analogs. For example in SbX_6^{-n} species, the bond lengths are 2.65 Å $(n = 3)$ and 2.35 Å $(n = 1)$ for the chloride and 2.80 Å $(n = 3)$ and 2.55 Å $(n = 1)$ for the bromide. This is a direct result of occupation of the a_{1g} AX antibonding orbital for the case where $n = 3$. The observation of distorted structures when O or F are present as ligands is usually ascribed to the "polarizing power" of these small ligands. We shall reach very similar conclusions as to the importance of s-orbital participation in influencing the stereochemistry of these molecules in the next chapter.

Independent of the question of how the structure is expected to change as the effect of the central atom s-orbital increases, is the importance of the latter on an absolute scale. There is no doubt that for the heavier elements of the right hand side of the periodic table there is a reluctance of the s^2 electron pair to enter into chemical bonding (note for example the stability of $Tl^{(I)}$ and $Pb^{(II)}$) and to influence molecular geometry (as noted in this section). For many years the origin of this inert pair effect was not understood but recently the effect of relativistic corrections on atomic wavefunctions has received some attention. As the nuclear charge increases, atomic orbital wavefunctions in general become more contracted (an increase in the exponent in the radial part of the wavefunction). This shrinkage affects s-orbitals much more than p-orbitals and the overall result for the heavy elements is contracted s-orbitals which have rather poor overlap with the orbitals on coordinated ligands.

6 Molecular Orbital Models of Main Group Stereochemistry: The Angular Overlap Approach

This approach to main group stereochemistry is at its most basic a quantified version of the p orbital-only model originally proposed by Rundle and Pimentel[172,182,183] to rationalize the long I–I bonds in I_3^- compared to I_2 and in other molecules where there are more than four electron pairs (an octet) around the central atom (hypervalent molecules).[152,153] We describe this approach and its application to these molecules but then show that with very little extension the structures of all main group AY_n molecules may also be described. The basic premises of the Rundle and Pimentel approach are twofold: (1) The outer d orbitals on the central atom may be neglected. Recall that in the hybridization scheme (Section 3.2) for more than four electron pairs, hybrids involving d orbitals need to be invoked. The necessity of their inclusion disappeared when using the molecular orbital model (this is not to say however that they do not contribute to bonding) and the structure of ClF_3 for example was produced naturally without them. (2) The central-atom s orbital is neglected except as a storage location for one pair of electrons. Since the s–p energy separation is often quite large this perhaps is not too drastic an approximation. The development of the scheme was prompted by the observation of a large number of 90 and 180° YAY bond angles in molecules which suggested that perpendicular orbitals (i.e., the three mutually perpendicular p orbitals) are predominantly used in A–Y bonding.

6.1   Two Types of Bond

For two AY bonds at right angles to one another each ligand σ orbital is able to interact with a different central-atom orbital. Using the angular overlap approach introduced in Section 2.6 we may derive a semiquantitative molecular orbital diagram of the form of Figure 6.1. The overlap integral of each ligand σ orbital with one p orbital is S_σ and the total stabilization energy for the system with two pairs of electrons in the lowest two orbitals is clearly $\Sigma(\sigma) = 4e_\sigma - 4f_\sigma$. For the case of two ligands sharing the same p orbital we need to write symmetry-adapted functions (Equations 6.1),

$$\psi_g = \frac{1}{\sqrt{2}}(\phi_1 + \phi_2)$$
$$\psi_u = \frac{1}{\sqrt{2}}(\phi_1 - \phi_2)$$

(6.1)

ψ_g is of the wrong symmetry to interact with the p orbital, but ψ_u has an overlap integral.

$$S_u = \frac{2}{\sqrt{2}} S_\sigma$$

(6.2)

such that the total stabilization energy of the bonding orbital is (Figure 6.2)

$$2\beta_\sigma S_\sigma{}^2 - 4\gamma_\sigma S_\sigma{}^4 = 2e_\sigma - 4f_\sigma$$

(6.3)

We could of course have reached the same conclusion using the ligand additivity Equation 2.35. With two pairs of electrons associated with the

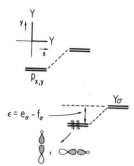

Figure 6.1   Energy level diagram on a p orbital only model using the angular overlap approach model for an AY_2 molecule with $YAY = 90°$ (σ orbitals only included on Y).

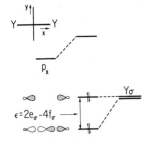

Figure 6.2 Energy level diagram on a p orbital- only model for a linear AY_2 molecular using the angular overlap approach (σ orbitals only included on Y).

p-orbital manifold, the stabilization energy of this configuration is simply $\Sigma(\sigma) = 4e_\sigma - 8f_\sigma$. Since e_σ and f_σ are both negative, the case where the two ligands share the same orbital is less stable than the one where the two ligands interact with two different orbitals. (The argument is identical to the one used for the allene molecule in Section 2.5 and for O_2H_2 and O_2F_2 in Section 4.3. If the total stabilization energy is divided into two, to find the contribution to each AY bond a stabilization energy of $2e_\sigma - 2f_\sigma$ is found for a normal $2c–2e$ bond (two center-two electron) and $2e_\sigma - 4f_\sigma$ for each $3c–4e$ bond. Already then we have a rationalization for the longer (weaker) bonds in I_3^- compared to I_2 itself. The scheme is readily extended to more complex molecules. For example in ClF_3 the shorter unique bond was a facet of the geometry which the VSEPR scheme handled by an assumption and the second-order Jahn–Teller approach handled by noting an admixture of e' stretching and bending distortions from the trigonal plane. We may envisage its occurrence in another way as arising via the scheme of Figure 6.3. One p orbital is shared by two ligand σ orbitals, and these bonds are weaker and hence longer than the bond which involves no p-orbital sharing. SF_4 may be viewed analogously. For the case of the trigonal bipyramidal AF_5 (e.g., PF_5) molecule (Figure 6.4) the axial linkages (interacting with p_z only) may be separated from the equatorial ones (involving p_x and p_y only). The derivation of the stabilization energy of the a_2'' orbital is exactly the same as for the linear case above. For the three equatorial ligands,

$$\psi_e'(1) = \frac{1}{\sqrt{6}}(2\phi_1 - \phi_2 - \phi_3)$$

$$\psi_e'(2) = \frac{1}{\sqrt{2}}(\phi_1 - \phi_2)$$

(6.4)

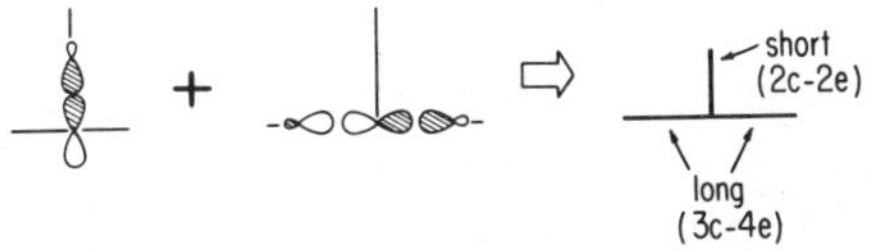

Figure 6.3 Rationalization of the relative bond lengths in ClF_3 and BrF_3.

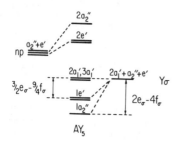

Figure 6.4 Energy level diagram for a trigonal bipyramidal AY_5 molecule using the angular overlap approach and the p orbital only model (σ orbitals only included on Y).

Recall that the overlap integral between a p orbital and a ligand σ orbital located at an angle θ to its axis is simply $S_\sigma \cos \theta$ (Table 1.1) so that $S_{e'} = \sqrt{3/2} S_\sigma$ and the stabilization energy of each e' symmetry component is $\frac{3}{2}e_\sigma - \frac{9}{4}f_\sigma$. Allowing for two electrons in each of the four lowest orbitals (remember one extra pair is in the As orbital) and dividing the stabilization energies associated with the axial and equatorial AY bonds by two and three respectively, the σ stabilization energies per AY linkage $[\Sigma_i(\sigma)]$ are

$$\Sigma_{ax}(\sigma) = 2e_\sigma - 4f_\sigma$$

$$\Sigma_{eq}(\sigma) = 2e_\sigma - 3f_\sigma \qquad (6.5)$$

Thus we expect the axial bonds to be longer (weaker) than the equatorial ones, which is indeed the experimental observation (Figure 3.4) for PX_5. The general result is clearly that the shorter (stronger) bonds are associated with those AY linkages where the ratio R = (number of central-atom p orbitals used)/(number of equivalent ligands) is maximized; so that whenever possible ligands avoid sharing a central-atom orbital. Interestingly our method shows (we develop it more fully in Chapter 12) that whereas $r_{ax}/r_{eq} > 1$ for a trigonal bipyramidal molecule such as PF_5 ($R_{ax} = \frac{1}{2}$, $R_{eq} = \frac{2}{3}$), for a pentagonal bipyramid such as IF_7 (Figure 3.4) the reverse is true ($R_{ax} = \frac{1}{2}$, $R_{eq} = \frac{2}{5}$) and $r_{ax}/r_{eq} < 1$.[200] This predicted trend on increasing coordination number is indeed found experimentally. Thus relative bond lengths in these AY_n molecules are readily tackled using this approach.

6.2 Generation of Walsh Diagrams

We proceed as before, ignoring the central-atom s orbital and use the AOM to quantify the diagrams developed in Chapter 4. Figure 6.5a shows

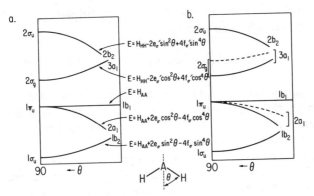

Figure 6.5 Orbital energy changes on bending a linear AH_2 molecule estimated using the AOM: (a) without the effect of the As orbital; this is just a plot against θ of the diagram of Figure 2.10; (b) As orbital interaction included showing the shifts of the a_1 orbitals (dashed lines).

the form of a σ-only, p orbital-only diagram for an AY_2 molecule. This is just a plot against θ of the energies calculated in Figure 2.10. The overlap of the ligand a_1 combination [$\psi(a_1) = 1/\sqrt{2}(\phi_1 + \phi_2)$] with the p_z orbital is $\sqrt{2}S_\sigma \cos\theta$, and the overlap of the ligand b_2 combination [$\psi(b_2) = 1/\sqrt{2}(\phi_1 - \phi_2)$] with p_y is simply $\sqrt{2}S_\sigma \sin\theta$. The Walsh diagram for AY_2 bending is readily generated following the ideas of Section 2.5. Addition of electrons allows calculation of the energy of a particular geometry. For BeY_2 the σ stabilization energy is

$$\Sigma(\sigma) = 2(2e_\sigma \sin^2\theta - 4f_\sigma \sin^4\theta) \tag{6.6}$$

Since e_σ is larger than f_σ the energy changes are dominated by the first term in parentheses which has a maximum at $\theta = 90°$. As before, molecules with this configuration, $(1b_2)^2$, are predicted to be linear. For singlet CY_2 or OY_2 $(1b_2)^2(2a_1)^2$, the stabilization energy is

$$\Sigma(\sigma) = 2(2e_\sigma - 4f_\sigma[\cos^4\theta + \sin^4\theta]) \tag{6.7}$$

The term in e_σ is now angle independent (as it should be under the AOM summation rule, Equation 2.34). The minimum value of the term in f_σ (since it is included with a minus sign) is readily found (by differentiation of $\cos^4\theta + \sin^4\theta$ with respect to θ) to be at $\theta = \sin^{-1}(1/\sqrt{2}) = 45°$, namely, when the YAY angle is $90°$. Thus bent singlet CY_2 and OY_2 species are predicted to be most stable, although for the first row systems the bond angles are somewhat larger than $90°$. For species with an unpaired electron in the $2a_1$ orbital BY_2 or CY_2 (triplet) the stabilization energy may be readily seen to be given by

$$\Sigma(\sigma) = 2e_\sigma[2 \sin^2 \theta + \cos^2 \theta] - 4f_\sigma[2 \sin^4 \theta + \cos^4 \theta] \qquad (6.8)$$

with a maximum in the e_σ term at $\theta = 90°$. These species should be linear. Both BH_2 and triplet CH_2 are however bent and the predictions of the simple scheme are in error. XeY_2 with two more electrons than OY_2 is certainly expected to be linear.

Analogous trigonometry leads to the diagram of Figure 6.6 for bending the planar AY_3 to a pyramidal structure. BY_3 systems are predicted to be planar ($\theta = 90°$), NY_3 species are predicted to be pyramidal with $\theta = \cos^{-1}\sqrt{1/3} = 54.7°$. This is the geometry with all AY bonds at right angles to one another, a result also suggested by an analysis using the orbital sharing ratio of the previous section. In the planar geometry for NY_3, $R = \frac{2}{3}$ and in the pyramidal $R = 1$, that is, the lowest energy angular geometry is where the nitrogen atom p orbitals share the smallest number of ligand σ orbitals. (We cannot use a similar approach for BY_3 since not all the bonding orbitals are occupied and the geometry is determined by the e_σ and not the f_σ term.)

6.3 The T-Shape Molecule and the Inclusion of the Central-Atom s Orbital

These AY_3 molecules have the option in addition of distorting to a T shape. Using analogous methods we may derive an energy correlation diagram for this distortion mode (Figure 6.7a). Note that an accidental degeneracy occurs for the lowest two orbitals at the arrowhead geometry with $\theta = 120°$. One feature of the diagram is that the middle a_1 orbital

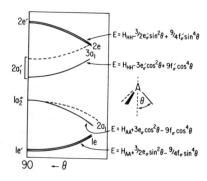

$E = H_{HH} - \frac{3}{2} e_\sigma' \sin^2\theta + \frac{9}{4} f_\sigma' \sin^4\theta$

$E = H_{HH} - 3 e_\sigma' \cos^2\theta + 9 f_\sigma' \cos^4\theta$

$E = H_{AA} + 3 e_\sigma \cos^2\theta - 9 f_\sigma \cos^4\theta$

$E = H_{AA} + \frac{3}{2} e_\sigma \sin^2\theta - \frac{9}{4} f_\sigma \sin^4\theta$

Figure 6.6 Energy level changes on pyramidalizing a planar AY_3 molecule using the angular overlap approach. Dashed lines show effect of As orbital mixing with a_1 orbitals.

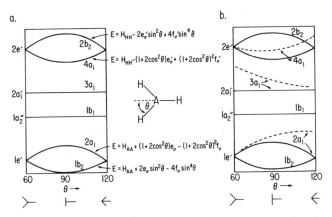

Figure 6.7 Energy level changes on distortion of a planar (D_{3h}) AY$_3$ molecule to a T shape: (*a*) without and (*b*) with *s* orbital mixing.

($3a_1$) of the distorted geometry remains unchanged in energy on distortion. This is purely a consequence of the fact that there are three (neglecting the As orbital) atomic a_1 orbitals, two located on the ligands and one on the central atom. Solution of the relevant secular determinant shows that one orbital remains unchanged in energy but the other two change as the interaction is switched on (recall the case of the allyl species in Section 2.3). With two electron pairs,

$$\Sigma(\sigma) = 6e_\sigma - 2f_\sigma[4 \sin^4 \theta + (1 + 2 \cos^2 \theta)^2] \qquad (6.9)$$

which has a maximum value (minimum value of the term in brackets) at $\theta = 60°$, namely, the D_{3h} geometry. Thus all three molecules BY$_3$ (two p manifold electron pairs), NY$_3$ (three pairs), and ClY$_3$ (four pairs) are all stable at the trigonal planar geometry and are not predicted to distort to the T shape on this p orbital-only scheme. The distortion of the ClY$_3$ molecule to a T shape is not predicted since $3a_1$ is not stabilized on bending.

At this stage we need to abandon our premise that only p orbitals control the stereodynamics. While this assumption worked well for many systems, for hypervalent molecules with more than four electron pairs the effect of s-orbital interaction needs to be included to reach the observed geometry. This may be done by modifying a diagram such as Figure 6.7a to illustrate the energy shifts of the molecular orbitals as the s-orbital interaction is switched on. Figure 6.8 shows the basic strategy. The central-atom s orbital is always of the highest possible symmetry species (a_1'

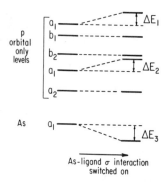

Figure 6.8 Strategy involved in including the central-atom s orbital in bonding showing interaction with orbitals of a_1 symmetry. $\Delta E_1 + \Delta E_2 \sim$ constant and $\Delta E_3 \sim$ constant as the system is distorted.

or a_1 in most of our examples). Thus when the p orbital-only set of molecular orbitals for the AY_n system is allowed to interact with this s orbital, only those of a_1 symmetry will be involved. The result of such a perturbation is to push these a_1 orbitals to higher energy. However the ligand sum rule of Equation 2.34 means that to a good approximation the total destabilization energy of all these p-manifold a_1 orbitals induced by the central-atom s orbital will be a constant irrespective of geometry. Similarly the total stabilization energy afforded the s orbital should be constant as a function of geometry, and only small changes in the energy of this orbital are in fact found in quantitative theoretical studies as the molecule is distorted (see Section 7.4).

At the D_{3h} geometry there is only one p-manifold a_1 orbital, and this receives all the latent destabilization energy via s-orbital interaction (Figure 6.7b). On distortion toward a T shape, two new a_1 orbitals are split off from the e' orbitals of the trigonal planar structure. As the molecule distorts and these orbitals get less and less like the parent e' orbitals of the D_{3h} geometry, so the s-orbital interaction with them increases. Since the total destabilization energy of all three a_1 orbitals ($2a_1$, $3a_1$, and $4a_1$) must remain constant, this means that the $3a_1$ orbital decreases in energy as the molecule distorts so that ClF_3, with two electrons in this orbital, may be stabilized by bending away from the D_{3h} geometry. Since $2a_1$ is destabilized at the T and arrowhead geometry, we have an additional explanation for the instability of the latter geometry for BY_3 and NY_3. (In the sharing out of the destabilization energy $2a_1$ will be destabilized less than $3a_1$ is stabilized on distortion to the T shape.) Compare Figure 6.7b with the quantitative result of Figure 4.7b. The addition of the As orbital to the Walsh diagram describing bending with the C_{3v} coordinate for AY_3 is shown in Figure 6.6. While its inclusion makes no difference to the geometry arguments for the three-pair (BY_3) case, as the s-orbital involvement increases, the geometries of four-pair molecules should become less pyramidal. This result is matched by quantitative calculations as described in Section 7.4.

We need to be a little careful in the derivation and use of these p orbital-only molecular orbital diagrams. The $2a_1'$ orbital, for example of the planar AH$_3$ molecule (Figure 6.6), is the out-of-phase combination of the As orbital with the H $1s$ orbitals. Its location on the p orbital-only molecular orbital diagram has been artificially maintained so that the level ordering is similar to that expected in the full molecular orbital diagram. Alternatively (as in Figures 2.10 and 6.6) the central atom may be assumed to be more electronegative than the ligands.

6.4 AY$_4$ Molecules

Three coordinate D_{3h} molecules have two distortion routes from this geometry—to a T shape and to a pyramid. A combination of both would give a C_s molecule which we did not discuss. Four coordinate molecules have even more possible avenues of distortion. We shall only investigate some of the higher-symmetry ones. The construction of a Walsh diagram for the $T_d \rightarrow C_{3v}$ distortion is a simple matter since this four-coordinate molecule may be viewed as a three-coordinate pyramidal species with an extra ligand added. (Recall the ligand additivity rule of the AOM of Section 2.5) Plotting out the orbital energies as a function of angle gives Figure 6.9. (Note that the $3a_1$ orbital remains unchanged in energy on the p

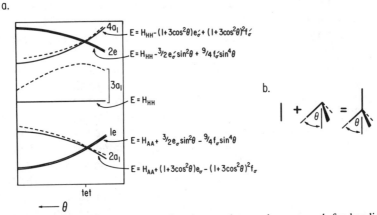

Figure 6.9 (*a*) Energy level diagram using the angular overlap approach for bending an AY$_4$ molecule within the C_{3v} coordinate. Dashed line indicates effect of s-orbital mixing. (*b*) Schematic derivation of the quadratic part of the energy level diagram from pyramidal AY$_3$ and linear AY units.

orbital-only model since it is the middle member of a three orbital set.) Inclusion of s-orbital interaction along the lines of the previous section leads to dashed lines for the a_1 orbitals. $3a_1$ receives all the destabilization energy at the T_d geometry and therefore must be stabilized on distortion.

For four electron pairs (CH_4) the total stabilization energy is

$$\Sigma(\sigma) = 8e_\sigma - 2f_\sigma \left[\frac{9}{2} \sin^4 \theta + (1 + 3 \cos^2 \theta)^2 \right] \qquad (6.10)$$

Minimization of the function in brackets gives a value of $\theta = \sin^{-1} \sqrt{8}/3$ for the most stable geometry. This is the tetrahedral structure observed experimentally for these species. In addition distortion away from the tetrahedron is discouraged by the destabilizing forces of As–ligand σ interaction. For five electron pairs the $3a_1$ orbital is doubly occupied and on the p orbital-only model does not change in energy on distortion. Inclusion of s-orbital interaction shows that SY_4 is clearly unstable at this geometry and should distort. There are geometries other than C_{3v} to consider as well, where the t_2 orbitals have split apart in energy and the $3a_1$ orbital is stabilized, which we look at shortly. XeF_4 is Jahn–Teller unstable at the tetrahedral geometry $(2t_2)^2$. Opening up of this geometry along the C_{3v} coordinate to $\theta > \theta_{tet}$ stabilizes the molecule and removes the degeneracy but still leaves the highest-energy electrons in an orbital which is antibonding between the central atom and ligands.

Derivation of molecular orbital diagrams for other distortion modes is straightforward (Figure 6.10). Given that the five electron-pair molecules SY_4 will not be tetrahedral, which one of these geometries will they adopt? For the two coordinates (C_{2v} and C_{4v} geometries) of Figure 6.10, minimization of the term in f_σ leads to the conclusion that the electronically favored geometry is the C_{4v} one with $\theta = 55°$, that is, where the trans YAY angles are 110° and the cis angles are about 75°. Ab initio molecular orbital calculations on SH_4 disagree as to the lowest energy arrangement. The C_{4v} and C_{2v} butterfly geometries are very close in energy. The C_{4v} structure is of course not the one favored using VSEPR arguments. Experimentally it has been found for a few species only. The PbO_4 unit in crystalline red PbO (Fig. 7.12) has bond angles of 75(4)° and 115(2)°, close to those predicted on the p orbital-only model. A similar Bi–O coordination occurs in several complex oxides containing the $Bi^{III}O_4$ unit. Tetragonal pyramids are also found in Pb^{II} diethylthiocarbamate and diethyldithiophosphate in the mineral aikinite[123] and interestingly in mackinawite, a form of FeS. We return to a discussion of this point in a later chapter but note here that the nonbonded distances in this C_{4v} geometry are very short and might lead to large nonbonded repulsions,

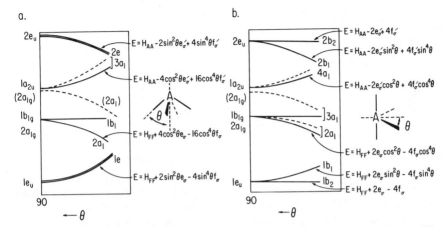

Figure 6.10 Energy level diagrams using the angular overlap approach for bending an AY_4 molecule within (*a*) the C_{4v} and (*b*) the C_{2v} distortion coordinate. Dashed lines indicate the effect of *s*-orbital interaction. In these diagrams we have made the ligands more electronegative than the central atom, e.g., XeF_4 and SF_4.

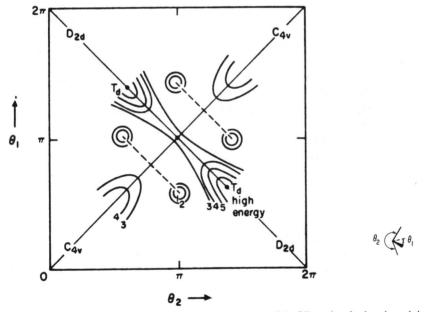

Figure 6.11 Potential energy surface for rearrangement of the SF_4 molecule showing minima at the butterfly (C_{2v}) geometry and saddle points at the C_{4v} structures. The four C_{2v} minima lie at about $\theta_{1,2} = 105$ and $180°$. The $\theta_1 = \theta_2$ diagonals describe C_{4v} geometries, and D_{2d} structures lie along the line $\theta_1 = 2\pi - \theta_2$. Dashed lines are Berry pseudorotations. Reprinted with permission from M. M. L. Chen and R. Hoffmann, *J. Am. Chem. Soc.* **98** 1651 (1976). Copyright by the American Chemical Society.

a.

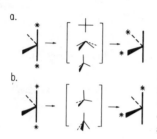

b.

Figure 6.12 Two permutationally distinct rearrangement modes of a butterfly SF_4 molecule. The asterisks represent site labels. Process (a) is compatible with transition states of D_{4h}, C_{4v}, or T_d symmetry. Process (b) is compatible with a transition state of C_{3v} or C_s symmetry.

a complaint that does not afflict the C_{2v} geometry. The energetic proximity of C_{2v} and C_{4v} geometries defines the way the axial and equatorial fluorine atoms exchange via a thermally induced intramolecular rearrangement process. Figure 6.11 shows a part of a calculated[37] potential energy surface for SF_4, and Figure 6.12 shows two ways the fluorine atoms could exchange. Analysis of DNMR (dynamic nuclear magnetic resonance) results [121] shows that the exchange occurs via the first of the two schemes consistent with our theoretically predicted C_{4v} transition state.

The advantage of the square planar geometry for XeF_4 is now clear from Figure 6.10a. The two electrons in the a_{2u} orbital of the planar geometry are in a nonbonding orbital (a pure p_z orbital on Xe). In all other geometries this orbital is destabilized by interaction with the ligand σ orbitals. However the $2a_{1g}$ orbital that may lie above or below this orbital (we assumed it lay above in Section 6.5) contains all the s-orbital destabilization energy at this geometry and is hence stabilized as soon as the geometry is distorted in any direction. Thus the square planar geometry seems to be favored on this model by the superiority of the p- rather than s-orbital effects. Recall a similar discussion in Section 5.2 where increasing s-orbital interaction is predicted to lead to a distorted structure.

6.5 AY_5 Molecules

Figure 6.14 is readily derived for the Berry[17] process (Figure 6.13) defined in general by the two angles θ_1 and θ_2. Here there are two nonbonding a_1 orbitals in the tbp on the p orbital-only model. We can immediately write down their wavefunctions (Equations 6.11):

$$\psi_1 = \frac{1}{\sqrt{5}}(\phi_1 + \phi_2 + \phi_3 + \phi_4 + \phi_5)$$

$$\psi_2 = \frac{1}{\sqrt{30}}[3\phi_1 + 3\phi_2 - 2(\phi_3 + \phi_4 + \phi_5)]$$

(6.11)

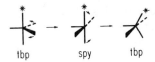

tbp   spy   tbp

Figure 6.13 Berry pseudorotation process linking trigonal bipyramidal and square pyramidal geometries. (Note exchange of axial and equatorial sites, labeled with an asterisk.)

These correlate with a_1 and b_1 orbitals on the spy. The a_1 orbital here is simply

$$\psi_3 = \frac{1}{2}(\phi_2 + \phi_3 + \phi_4 + \phi_5) \qquad (6.12)$$

For all systems where the lowest three orbitals of Figure 6.14 are occupied (four valence electron pairs), the energy term in e_σ is of course angle independent (Equation 6.13):

$$\Sigma(\sigma) = 10e_\sigma - f_\sigma[4\sin^4\theta_1 + 4\sin^4\theta_2 \\ + (1 + 2\cos^2\theta_1 + 2\cos^2\theta_2)^2] \qquad (6.13)$$

The f_σ term is minimized for $\theta_1 = 90°$, $\theta_2 = 60°$, that is, at the tbp geometry. On this basis then, since the two next highest orbitals are unchanged in energy across the tbp–spy interconversion coordinate, the tbp geometry is expected for five and six valence-pair molecules on the p orbital-only scheme. The destabilization energies of ψ_{1-3} on switching on the s-orbital interaction with the a_1 orbitals in spy and tbp geometries will be proportional to the squares of the overlap integrals of these functions with the central-atom s orbital. These energies can readily be seen to be in the ratio of $5:0:4$ for ψ_{1-3}. This leads to the dashed lines in the diagram and shows that the $3a_1$ orbital is stabilized on distortion from tbp to spy, which

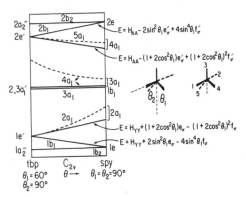

Figure 6.14 Energy level diagrams using the angular overlap approach for distortion of an AY₅ molecule in a C_{2v} coordinate linking square pyramidal and trigonal bipyramidal geometries. Dashed lines indicate energy shifts of a_1 orbitals on including the central-atom s orbital. (The quadratic energy dependence on the angles θ_1 and θ_2 in general are a_1; $(1 + 2\cos^2\theta_1 + 2\cos^2\theta_2)e_\sigma$, b_1; $2\sin^2\theta_1 e_\sigma$, b_2; $2\sin^2\theta_2 e_\sigma$).

immediately suggests that BrF_5 with six valence pairs should be a square pyramid. The $2a_1$ orbital of the spy also receives some destabilization at this geometry. This orbital is heavily involved in interaction with the Ap_z orbital and will contain a smaller amount of ligand σ character than ψ_{1-3}. Its energetic changes on distortion will be smaller than for the higher-energy a_1 orbital. Its slope however is in the right direction to help stabilize the tbp geometry for the five electron-pair case.

6.6 AY$_6$ Molecules

The six electron-pair case SF_6 is readily understood on the basis of our previous arguments. The best stabilization of the three p orbitals is as the degenerate t_{1u} set of an octahedral geometry. With this electronic configuration, the term in e_σ is angle independent, but the f_σ term is minimized here on the basis of the orbital-sharing ideas. The $2a_{1g}$ orbital (unoccupied) receives all the latent s-orbital destabilization energy at this geometry. On distortion of the molecule so that this octahedral arrangement is destroyed the t_{1u} set will split apart in energy. In several point groups an a_1 orbital results. This now receives a part of the s-orbital destabilization energy (Figure 6.15). SF_6 may be regarded as being fixed in this octahedral geometry because any other would result in strong destabilization of a_1 orbitals by s-orbital interaction and a loss in p-orbital stabilization energy as a result of a less favorable f_σ coefficient. For XeF_6 with seven pairs, distortion from octahedral is strongly encouraged by occupation of the $2a_{1g}$ orbital. How the molecule distorts is a question not readily answered on the scheme since there are several routes open to us. It will be a balance of the stabilization energy of $2a_{1g}$ as it becomes less As–ligand σ antibonding and the destabilization of the lower orbitals as the f_σ coefficient becomes larger.

On the p orbital-only scheme, $2a_{1g}$ will remain equienergetic on distortion and XeF_6 and related molecules should be undistorted. This conclusion is often referred to as "the molecular orbital prediction" of the XeF_6 geometry, but as we have seen it is only one way of looking at the

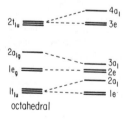

Figure 6.15 Effect of distorting octahedral AY$_6$ species showing destabilization of a_1 orbitals. Labels used correspond to a C_{3v} structure.

picture. As ligand σ–central-atom s orbital interaction is switched on, the tendency of the molecule to distort increases, an argument exactly analogous to the one we have just used for XeF_4 in this chapter and also on the second-order Jahn–Teller model of Sections 5.5 and 5.9.

6.7 The Factors Determining Molecular Geometry

A useful summary of this blow-by-blow description of the application of the AOM to main group stereochemistry is given in Table 6.1. In each case where we have suggested that it is the inclusion of the central-atom s orbital that has determined the geometry, we assumed that the relative sizes of the energetic contributions were such that the preferences of the f_σ part of $\Sigma(\sigma)$ were of secondary importance. In several molecules labeled with a d, the inclusion of the s-orbital interaction helped consolidate the geometries found on the p orbital-only model. For XeF_4 (and also in fact for XeF_2) we assumed that the e_σ part of the p orbital forces were energetically superior to the s orbital forces in determining the geometry. This relative order of importance in determining the geometries of these molecules,

$$e_\sigma(p) \text{ term} > sp \text{ mixing} > f_\sigma(p) \text{ term}$$

is then quite a useful one and rationalizes the apparent discrepancy found in the second-order Jahn–Teller discussion of the species XeF_4 and XeF_6.

Table 6.1 Factors Determining Molecular Geometry

Number of Ligands	Number of Electron Pairs					
	2	3	4	5	6	7
2	$BeF_2{}^a$	$CF_2{}^{b,e}$	$OF_2{}^{b,e}$	$XeF_2{}^a$		
3		$BF_3{}^a$	$NF_3{}^{b,e}$	$ClF_3{}^c$		
4			$CF_4{}^{b,d}$	$SF_4{}^c$	$XeF_4{}^a$	
5				$PF_5{}^{b,d}$	$BrF_5{}^c$	
6					$SF_6{}^{b,d}$	$XeF_6{}^c$
7						IF_7

[a] e_σ terms determines geometry.
[b] f_σ terms determines geometry.
[c] s–p orbital interaction distorts geometry from symmetrical structure.
[d] s–p orbital interaction holds geometry at symmetrical structure.
[e] s–p orbital interaction compatible with this geometry.

Both were predicted to be distorted from their symmetric structures by sizable s-orbital involvement. Here we see that in the XeF_6 case the s-orbital forces need only to overcome energy terms in f_σ on distortion, but in the square planar XeF_4 case energy terms in e_σ must be overcome. (Recall that for XeF_4 in all geometries other than square planar two electrons occupied an Ap–ligand σ antibonding orbital.) A glance at Table 6.1 also shows why the p orbital-only model is successful for a variety of systems. Either, as in BeF_2 or BF_3, s-orbital interaction has no influence on that part of the orbital diagram or, as in CF_4, s- and p-orbital forces work hand in hand to stabilize one particular geometry.

7 Further Comments on the Structures of Main Group Molecules

Until now we have ignored steric interactions between the ligands and have been able to understand a large number of structural results purely by consideration of electronic (usually σ bonding) forces. We noted in Section 2.6 that there was a molecular orbital way to regard nonbonded repulsions. In general we shall not be too concerned about the physical origin of these effects but more with their structural manifestation. In most cases we shall find perhaps rather subtle effects on molecular geometry, but in a few instances inclusion of nonbonded considerations allows a rationalization of gross structural effects. In several instances the results obtained by the use of steric arguments are very similar to those obtained by including π bonding. In this chapter we discuss both of these topics, which at first sight appear unrelated, together. In addition we investigate the specific problem of ligand site preferences in main group compounds and the relationship of quantitative calculations (varying in "quality") to the simple models we have described in previous chapters.

7.1 Ligand Site Preferences

If the geometric arrangement of ligands in an AY_n molecule is such that there are two symmetry-unrelated sets of AY linkages, how will the ligands Y and Y' distribute themselves between the two sites in the $AY_{n1}Y'_{n2}$ ($n = n_1 + n_2$) molecule? Several factors will be important. Bulky substituents will prefer the less crowded coordination positions and electronic factors, σ and π bonding may preferentially stabilize one particular arrangement of the Y and Y' ligands.

The distribution of electron density in the $3c-4e$ bonding situation which we described for systems such as ClF_3, SF_4, PF_5, is, in molecular orbital terms, analogous to that of the allyl negative ion, approached from the Hückel viewpoint in Section 2.3. The charge is higher on the end atoms of the three-center system than on the middle atom. In addition the negative charge is larger on these end atoms than found in the $2c-2e$ situation (e.g., ethylene). So in all of the examples containing inequivalent sets of ligands in Chapter 6, firstly the three-center bonds should be weaker and secondly the end atoms (ligands) should carry the highest charge. Some results from quantitative EHMO calculations on molecules of this type are shown in Figure 7.1, which bear out this simple analysis.

The more electronegative ligands (out of Y and Y') will search out those sites where the potential electron density is highest. This will result in the most stable arrangement since this large electron density will be associated with the deep-lying orbitals of the electronegative ligand. Some examples illustrating this charge control over the ligand site preference in molecules of this type are shown in Figure 7.2. For the trigonal bipyramid the three equatorial ligands are bound via two central-atom p orbitals. The charge on these ligands is expected to lie between that expected for normal $2c-2e$ and $3c-4e$ bonds. The higher charges on the terminal compared to central atoms underlies the general rule that the most electronegative ligands are found in the former sites. Thus ClF_3 contains a central Cl atom Cl(—F)$_3$ rather than Cl—F(—F)$_2$. Although NF_2 is only found as FNF, OCl_2, which contains atoms of similar electronegativity, is known in two forms, as ClOCl and the unstable OClCl. N_2O is only found as NNO.

The limited number of van der Waals molecules that have been studied obey similar rules although it is not clear why they should. Thus linear

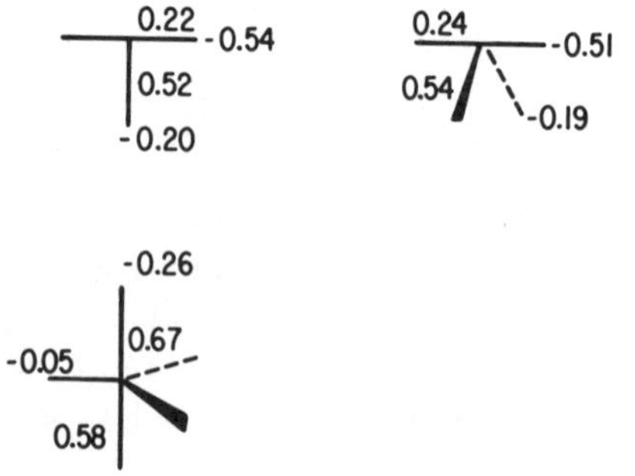

Figure 7.1   EHMO results on a series of main group molecules showing AY bond overlap populations and ligand charges ClH_3, SH_4, and PH_5.

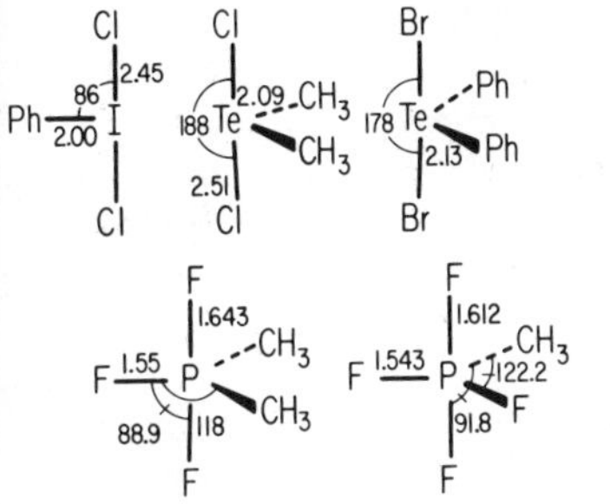

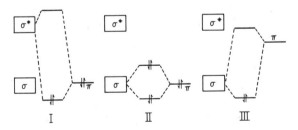

Figure 7.2 Preferential location of electronegative ligands in sites of highest charge (see Figure 7.1).

Ar . . . ClF contains two "coordinate" chlorine (the molecule is isoelectronic with XeF_2); Ar . . . CO_2 is a T-shaped molecule (isoelectronic with NO_2Cl); and Ar . . . BF_3 ("isoelectronic" with BF_4^-) is a trigonal pyramid. In each case the Ar is coordinated to the least electronegative center.

We may also inquire where π donors and acceptors prefer to reside independent of any considerations concerning their electronegativity. Three types of interaction of the ligand π orbitals with the σ set of AY_n orbitals may be considered (Figure 7.3). Clearly the interaction I of a ligand π donor orbital with the σ^* set results in a stabilization of the system. This type of interaction however is usually unimportant because of the large $\sigma^*-\pi$ energy separation. The interaction II leads overall to a destabilizing effect simply because the antibonding orbital of the interaction is destabilized more than the bonding orbital is stabilized and both are doubly occupied. By way of contrast the interaction III results in an electronic stabilization of the system. For D_{3h} trigonal bipyramidal PH_5[95] there are orbitals of symmetry species e' and a_2'' which may interact with ligand π type orbitals.

Figure 7.4 shows how three types of interaction lead to favorable li-

Figure 7.3 Schematic illustration of three types of A–Y π interaction. I and II are for π donors and III is for π acceptors. The blocks labeled σ and σ^* represent the high-energy (AY σ antibonding) and low-energy (AY σ bonding and nonbonding) orbitals.

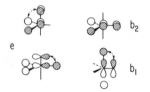

e

b₂

b₁

Figure 7.4 π interaction in the trigonal bipyramidal AY₅ molecule. Better equatorial interaction is found for the b_1 than the b_2 arrangement. The e-species interactions are of course equivalent.

gand–ligand overlap but how one equatorial interaction b_2 is less favored than the rest. By making use of the ideas of Figure 7.3, if the ligands are π acceptors then the axial sites will be preferred; if the ligands are π donors then the equatorial sites will be best. Since one of the equatorial π interactions (b_2) is less energetic than the other (b_1), a single-faced π acceptor will prefer (if in an equatorial site) to make use of the larger b_1 stabilization and a π donor will prefer the smaller destabilization associated with the b_2 arrangement. This is nicely borne out for the molecules of Figure 7.5 where for the thiotetrafluorophosphoranes low-temperature NMR results show nonequivalence of the axial fluorine atoms. (We can reach a similar result, equatorial π interaction larger than axial, by using a similar method to the one used for assessing relative σ-bond stabilization energies in these systems in Section 6.1. The π-acceptor stabilization energies are $\Sigma_{ax}(\pi) = 4e_\pi - 14f_\pi$ and $\Sigma_{eq}(\pi) = 4e_\pi - 13f_\pi$; but although this leads to the same structural preferences as found above, the numbers are of rather similar sizes.) In some systems the σ electronegativity and π effects compete. We predicted that π donors (such as fluorine) prefer the equatorial sites but electronegative ligands (such as fluorine) prefer the axial sites. In practice in these systems the σ effect appears to win out.

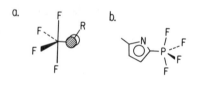

Figure 7.5 (*a*) Arrangement of π-donor SR unit in thiotetrafluorophosphoranes. (*b*) Analogous arrangement in 2-methyl-5-tetrafluorophosphoranylpyrrole.

7.2 π Bonding and Geometry

We have focused on σ interactions in describing molecular geometries, but it is important to see where the inclusion of π bonding might be significant. Let us take the four-valence pair AY₂ and AY₃ molecules. They are predicted on the basis of VSEPR or any of the molecular orbital

Figure 7.6 π-Type interaction in AY_2 and AY_3 systems which is maximized at the linear and trigonal planar geometries.

models to be bent or pyramidal, respectively. Yet some of these AY_2 species are either linear or have bond angles somewhat larger than expected and several AY_3 systems are planar. One argument which has been extensively used to rationalize the planarity of for example $C(CN)_3^-$ or $N(SiH_3)_3$ is to allow the π stabilization of the filled p_z orbital on the central atom by higher-energy unoccupied ligand orbitals to tip the scale in favor of this structure (Figure 7.6). Figure 7.7 shows that the π-type interaction in the AY_3 case, which has the smallest energy separation and so from perturbation theory might be energetically most important, is with the a_2'' orbital, a nonbonding orbital located entirely on the A atom on a σ-only model. The π interaction of species e' has a larger energy gap. Also the lower (filled) member of the interacting e' pair is diluted with ligand character as a result of σ interaction. Thus the π interaction of species a_2'' should be larger and is a maximum at the planar geometry (Figure 7.7). Note that the π-type orbitals on the ligands need to be of high energy and unoccupied for this mechanism to stabilize the planar geometry for four-pair systems. For the CN ligand an obvious candidate is the CN π* orbital. For SiH_3 the low-lying $2e'$ σ* orbital (Figure 4.6) is of the correct symmetry; $3d$ orbitals often suggested in the past for this purpose are not necessary. For ligands with filled low-energy π orbitals such as Y = halide (Figure 7.3) the overall result is a destabilizing one since the antibonding orbital is destabilized more than the bonding orbital is stabilized and these species will avoid the planar geometry. For BY_3 species (three pairs) and perhaps also for CY_3, π bonding will stabilize the planar structure. In the $P(SiH_3)_3$ species the longer P–Si distances may reduce the

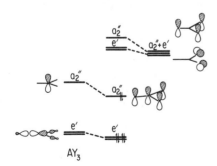

Figure 7.7 Molecular orbital diagram showing stabilization by π-acceptor orbitals on Y in planar AY_3. Only π orbitals of species a_2'' and e' are shown; those of species a_2' and e'' do not interact with any A-atom orbitals. Also only one component of the e' species orbitals is shown.

Table 7.1   Some Bond Lengths in Cl-
and SiH$_3$-Containing Systems (Å)

System	Bond Length
CH$_3$—SiH$_3$	1.867
CH$_3$—Cl	1.781
N—(SiH$_3$)$_3$	1.735 (planar)
N—Cl$_3$	1.759 (pyram)
P—(SiH$_3$)$_3$	2.248 (pyram)
P—Cl$_3$	2.043 (pyram)

importance of π bonding and the molecule is correspondingly pyramidal. In the isoelectronic O(HgCl)$_3^+$ species which is planar, the accessible low-energy orbitals on the Hg atom may be the $6p$ orbitals involved in π-type interaction with the chlorine atoms.

Evidence for π-type interaction in systems of this sort is found in shorter than expected AY bonds. Table 7.1 shows a comparison between the isoelectronic species SiH$_3$ and Cl. Clearly the N–Si distance is shorter than expected in N(SiH$_3$)$_3$, but no such shortening is found for P(SiH$_3$)$_3$. Planar N(SCF$_3$)$_3$ has[12] an N–S bond length of 1.705(2) Å, which is shorter than the N–Cl bond in pyramidal NCl$_3$. For two related systems where π bonding is not expected to be important, the bond to sulfur is the longer one. The C–Cl bond (1.781(1) Å) in CH$_3$Cl is shorter than the C–S bond (1.802(2) Å) in SMe$_2$. In N(SCF$_3$)$_3$ with a planar NS$_3$ skeleton the sum of the nonbonded hard-sphere radii from Table 7.2 is 2.90 Å, compared to the observed distance of 2.934 Å, so that on a nonbonded model often used (see next section) to rationalize these geometries the molecule is not sterically strained at the planar geometry; π bonding is probably the answer here.

Table 7.2   Rotational Barriers

Bond	Rotational Barrier (kJ/mole)	
	Calcd.	Obsd.
CH$_3$—CH$_3$	12.6	~11.7–12.6
CH$_3$—CH$_2$CH$_3$	13.8	~13.0–13.8
CH$_3$—C(CH$_3$)$_3$	17.2	~18.0
CH$_3$CH$_2$—CH$_2$CH$_3$	14.9	17.2 ± 1.7
	21.5	. . .
(CH$_3$)$_3$C—C(CH$_3$)$_3$	35.1	20.9?

Adapted from Ref. 10.

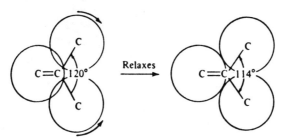

Figure 7.8 Relaxation of the bond angles in $CH_2{=}C(CH_3)_2$ on the hard-sphere model. Reprinted with permission from L. S. Bartell, *J. Chem. Educ.* **45** 754 (1968).

7.3 Nonbonded Radii

Bartell[10,11] noted that the CCC bond angles involving the bulky methyl groups in $CH_2{=}C(CH_3)_2$ were smaller than the 120° expected on the basis of simple traditional bonding ideas. If all the nonbonded carbon atoms were regarded as being sterically equivalent, then the geometry may be understood by defining a "hard sphere" radius for carbon which allows the 120° structure to relax (Figure 7.8). The CCC angle of 114° then reflects the fact that the C=C bond length is shorter than the C—C bond lengths. Extension of the scheme to other molecules of the ethylene type $YY'C{=}CY''Y'''$ and $O{=}CYY'$ led to the evaluation of a set of "hard sphere" radii for H, C, N, O, F, Cl, and Br able to reproduce the observed bond angles in these molecules with good accuracy (Figure 7.9). There are in fact small deviations in the directions expected on the electronegativity arguments (rule 3) of the VSEPR scheme (Section 3.1). In addition quantitative changes in bond lengths could be matched by using the relevant bond stretching force constants and a set of nonbonded potential energy functions adopted from the isoelectronic inert gas. The barriers to rotation in some of these hydrocarbons could also be predicted quite well (Table 7.2) by using this hard-sphere approach.

In some ways the nonbonded atom viewpoint where the geometry is determined by the packing of atoms around a central atom is phrased in similar terms to the VSEPR method with its packing of electron pairs. Thus the C=C bond is shorter than the C—C bonds in $H_2C{=}C(CH_3)_2$, and so the alkyl CCC angle is less than 120° on the nonbonded atomic model (Figure 7.8). On the VSEPR scheme the σ bonding pair of the multiple bond occupies more space around the coordination shell (rule 4). Similarly electronegative ligands have smaller nonbonded radii (a high effective nuclear charge contracts the valence electrons) in the hardsphere model. On the VSEPR scheme, electron pairs on the central atom

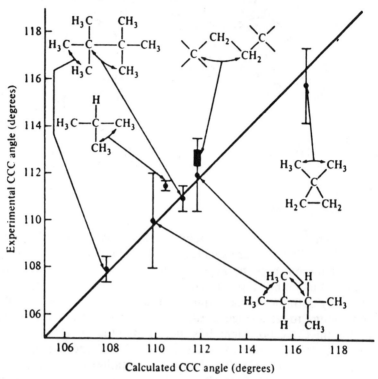

Figure 7.9 Calculated and observed bond angles in some hydrocarbons using the hard-sphere model. The solid block represents a range of angles in *n*-alkanes. Reprinted with permission from L. S. Bartell, *J. Chem. Educ.* **45** 754 (1968).

attached to an electronegative atom occupy less space in the coordination sphere. The two schemes are intrinsically different however. NF_3 for example is pyramidal, an observation which has a ready explanation on VSEPR grounds but not on the nonbonded scheme.

Glidewell[77] has extended this approach, and Table 7.3 shows a set of Bartell–Glidewell nonbonded hard-sphere radii for a range of main group elements. For an observed molecular geometry, bond angles, and bond lengths, the nonbonded distances are calculated. These may then be compared with the sum of the relevant hard-sphere radii. If the actual distance is less than this calculated strain-free one, then the geometry is considered to be sterically strained. If the geometry is an unexpected one, then steric forces could be responsible. For example the $C(CN)_3^-$ ion is planar and approximately trigonal in many salts. A pyramidal geometry is expected for such a molecule with four valence pairs, AB_3E. The actual CC nonbonded distances are 2.42 and 2.44 Å (twice), which is less than the sum

Table 7.3 Bartell–Glidewell Hard-Sphere Radii (Å)

					H
					0.92
Be	B	C	N	O	F
1.39	1.33	1.25	1.14	1.13	1.08
	Al	Si	P	S	Cl
	1.85	1.55	1.46	1.45	1.44
		Ge		Se	
		1.58		1.58	
		Sn	Sb	Te	
		1.88	1.88	1.87	

of two carbon radii from Table 7.3 of 2.50 Å. It is suggested that the structure is sterically strained. The pyramidal geometry is disfavored since it would increase the steric crowding by decreasing the CC non-bonded distances assuming the bonded CC distances remain constant.

Another species open to similar treatment is the $N(SiH_3)_3$ molecule, which has a planar NSi_3 skeleton. The isoelectronic species $N(CH_3)_3$ however is pyramidal in accordance with the VSEPR scheme. The actual Si–Si distances are 2.99 Å, which is shorter than the sum of the hard-sphere radii of 3.10 Å from Table 7.3. Again pyramidalization is not favored on the scheme since here an increase in steric strain is expected if the bond lengths remain fixed. Interestingly the infinite lattice of Si_3N_4 (in both α and β forms) contains N atoms in trigonal planar coordination by Si (Figure 7.10). The N–Si distances in these units are such that the Si–Si distances vary from 2.99 to 3.04 Å. Similar features are seen in Ge_3N_4. $P(SiH_3)_3$ is however a pyramidal molecule. The longer P–Si (compared to N–Si) linkages allow a sterically strain-free situation at the pyramidal geometry. A similar analysis of $N(CH_3)_3$ (Figure 7.11) shows that it is as sterically strained at the pyramidal structure as $N(SiH_3)_3$ is at the planar geometry. The percentage reductions in actual nonbonded distances compared to the sum of the relevant hard-sphere radii in the carbon and silicon analogs are the same (~4%). The important difference is that the $N(SiH_3)_3$ species would have a much larger steric strain if pyramidal. Similar arguments to those used for $N(SiH_3)_3$ may be used for other molecules. For triatomic skeletons analogous ideas apply to $(SiH_3)_2O$ (angle 144.1°), $(SiF_3)_2O$ (155.7°), $(GeH_3)_2O$ (126.4°), and $(SiCl_3)_2O$ (146°) where

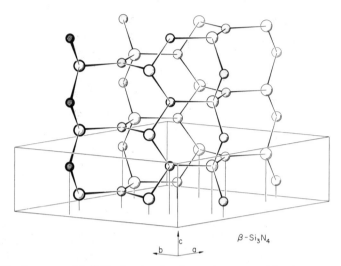

$\beta - Si_3N_4$

Figure 7.10   Structure of β-Si₃N₄ showing the planar nitrogen coordination. Reprinted with permission from *The Crystal Chemistry and Physics of Metals and Alloys* by W. B. Pearson, Wiley, New York (1972).

angles somewhat larger than their hydride (OH₂) or fluoro (OF₂) analogs are found. For the N(SCF₃)₃ system of the previous section the π-bonding argument appears the best since the molecule does not seem to be sterically strained by numerical evaluation of the nonbonded distances. For some of the other sterically strained systems however both π bonding and steric controls may well determine the actual geometry.

Interestingly we can apply the nonbonded arguments to the four electron-pair AH₂ and AH₃ molecules in Table 7.4. Recall that the VSEPR geometry predicted HAH bond angles of about 110°, and slightly smaller angles were indeed observed for the first-row species NH₃ and OH₂. Much smaller angles (termed anomalous on the VSEPR scheme) were however found for their heavier analogs. The *p* orbital-only model by way of contrast predicted angles of 90° for these systems. Table 7.4 shows the HH nonbonded distances observed in these hydrides and what they would be at the 90° geometry if the bond lengths remained unchanged. As may readily be seen, the heavier hydrides, because of their longer AH bonds, experience no strain at the 90° geometry and so electronic controls may win out; but the lighter hydrides are sterically strained at this structure

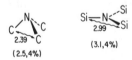

Si⟶N⟵Si
2.99 ⟍Si

(3.1,4%)

Figure 7.11   Nonbonded distances in pyramidal N(CH₃)₃ and planar N(SiH₃)₃ with the sum of the hard-sphere radii and the percentage difference in parentheses.

Table 7.4 Comparison of Bonded and Nonbonded Distances in AH_2 and AH_3 Systems

Species	Bond Length (Å)[a]	Observed Bond Angle (°)[a]	H–H Distance for 90° Angle	Actual H–H Distance
NH_2	1.024	103.4	1.448	1.607
PH_2	1.428	91.5	2.0195	2.044
OH_2	0.956	105.2	1.352	1.519
SH_2	1.382	92.2	1.878	1.914
BH_2	1.18	131	1.669	2.148
CH_2[b]	1.078	136[c]	1.525	1.999
NH_3	1.0173	107.8	1.439	1.644
PH_3	1.421	93.3	2.010	2.068
CH_3	1.079	120	1.526	1.869

[a] Data from G. Herzberg, *Electronic Spectra of Polyatomic Molecules,* Van Nostrand, New York (1966).
[b] Triplet.
[c] Original results suggested a linear species; newer data show a bent molecule.

and the observation of larger HAH angles may be a direct result of this. BH_2 and triplet CH_2 are interesting halfway molecules. They have half an electron pair on the VSEPR scheme or only one electron in an orbital rapidly stabilized on bending on an MO approach. Table 7.4 shows that they could clearly bend further if needed on electronic grounds, so the observed bond angle is not set by steric restraints. Our analysis shows that even at the 90° geometry the molecules are not sterically strained.

Another case where the steric argument is a particularly interesting one is in the geometry of four coordinate, five valence-pair molecules, for example SF_4 and the unknown SH_4 species. Recall that in Section 6.4 we concluded that the C_{4v} geometry was the most stable configuration for these five-pair species on electronic grounds (Figure 7.12). At the C_{4v} geometry with $\theta = 55°$ the smallest FF nonbonded distances are 1.82 Å, but the sum of the hard-sphere radii is 2.16 Å—a reduction of 16%. Here there are two pairs of ligand atoms linked by central-atom angles of 110°

2.3 75°
115°

a.

b.

Figure 7.12 Coordination of Pb in red PbO. Full circles are oxygen atoms: (*a*) local geometry; (*b*) solid-state environment showing four short and four long Pb–O distances.

and four pairs linked by angles of 75°. In the C_{2v} butterfly geometry there are five pairs of ligand atoms linked by angles of 90°, clearly less sterically strained with the smallest FF distance of 2.12 Å. Using the observed bond length in SH_2, the closest contact in the C_{4v} form of SH_4 is 1.68 Å, only a 9% reduction over the close contact limit. If a more realistic length of 1.45 Å is used for the SH distance to allow for the increase in coordination number, the reduction is only 4%. Thus the characteristic C_{2v} structure may occur by default electronically because the C_{4v} arrangement is usually so sterically strained. The C_{4v} AO_4 (Figure 7.12) structure in red PbO (the yellow form is a distorted version) is one of a small number of systems containing heavy central atoms with this geometry. To minimize ligand–ligand repulsions, long AY bonds are needed (found for the heavier atoms A) and small ligands (Y). Both criteria are met here. PbS however has the rock salt (NaCl) structure where each Pb atom is octahedrally coordinated.

Coordination numbers greater than four are rare for first-row (Li–Ne) elements. In CF_4 the FF, nonbonded distance is 2.15 Å (close to the hard-sphere contact limit of 2.16 Å). To maintain this in a trigonal bipyramidal CF_5^- species, assuming equal axial and equatorial bond lengths, requires stretching of the CF bonds from 1.317 Å to about 1.52 Å, a 16% increase. A similar factor applies to the chloride and bromide. In SiF_6^{-2} the A–F bonds are about 11% longer than SiF_4, and in GeF_6^{-2} they are about 3% longer than in GeF_4. For second-row atoms a similar calculation (assuming a pentagonal bipyramidal geometry AF_7) shows a 1, 7, 12, and 17% increase in AF bond length on increasing the coordination numbers of AlF_6^{-3}, SiF_6^{-2}, PF_6^-, and SF_6, respectively. This suggests that there is no steric prohibition of the synthesis of the (presently unknown) AlF_7^{-4} and SiF_7^{-3} species.

Increasing the coordination number of SF_6 or the CY_4 species results then in a switching on of considerable steric strain which may be relieved by the energetically unfavorable solution of stretching the AX bond. This observation may account for the rarity of coordination numbers of greater than four or six in first- and second-row systems. It also lends a more quantitative weight to the usual steric rationalization that although thermodynamically unstable with respect to hydrolysis, SF_6 and CX_4 species are quite kinetically inert to such a reaction. By way of contrast the heavier species SeF_6 and SiX_4 are much more reactive.

Steric control of coordination number may also be important in determining the structures of solids.[155] Many binary AY systems adopt the zincblende or wurtzite structure (4 coordination of A and Y), rocksalt (6:6 coordination), or cesium chloride (8:8 coordination) structures. Obviously the CsCl structure will not be favored for small cations A. It is found only for large species. Eight coordination is quite common in lanthanide and actinide compounds, where the metal atom is large.[146]

7.4   Results from Quantitative Calculations

The detailed dissection of the results of quantitative calculations may usefully be tied into the geometric models of previous chapters. Several sets of EHMO calculations have been performed on these main group molecules. The general form of the orbital energies on distortion are as we have described in previous chapters in qualitative terms. Quantitative geometry predictions are sometimes somewhat parameter sensitive (H_{ii} values of the atomic basis) unless the geometry is a symmetrical one (e.g., tetrahedral, square planar), but the direction of the distortion is usually correctly predicted for a realistic parameter choice. Varying the ns–np energy separation in EHMO calculations is a useful way of determining the importance of ns orbital participation in determining the angular geometry. As could be anticipated by our discussion in Chapter 6 it is found to be of crucial importance.[83,84] Figure 7.13 shows calculated Walsh dia-

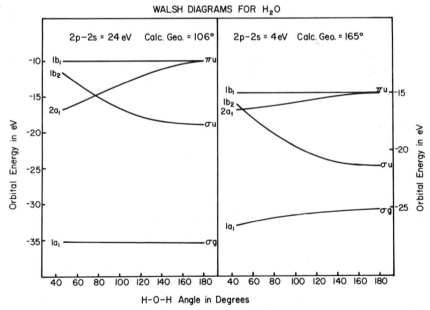

Figure 7.13   Walsh diagrams for the H_2O molecule with very different np–ns energy separations calculated using the EHMO method. Reprinted with permission from M. B. Hall, *Inorg. Chem.* **17** 2264 (1978). Copyright by the American Chemical Society.

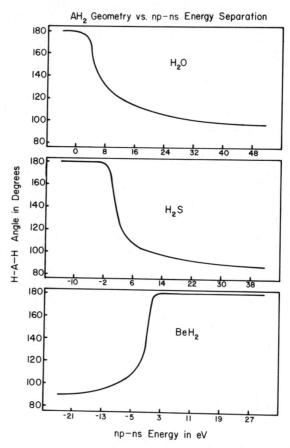

Figure 7.14   Calculated bond angles for some AH_2 systems as a function of np–ns energy separation. Reprinted with permission from M. B. Hall, *Inorg. Chem.* **17** 2264 (1978). Copyright by the American Chemical Society.

grams for an AH_2 system with two different values of the *ns–np* energy separation. With a large separation (actually, realistic H_{ii} values for O and H), the HOH angle in water is calculated to be 106° (observed 104.5°); but with a smaller separation (unrealistic H_{ii} values for water), the angle is calculated to be 165° (Figure 7.14). In Section 6.3 we suggested via perturbation theory considerations that the energy of the (mainly) 2*s* orbital of A remained unchanged in energy as the geometry changed. We can see that this is a perfectly reasonable assumption for the usual case of a large *ns–np* energy separation but not so when this gap is small (and when perturbation approaches may not be as valid). Indeed for the latter

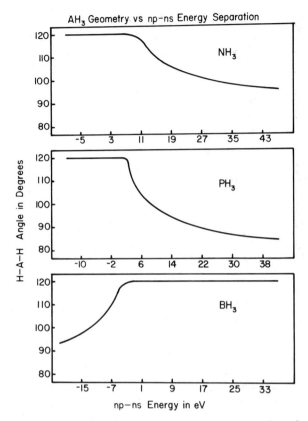

Figure 7.15   Calculated geometry of some AH_3 molecules as a function of np–ns energy separation. Reprinted with permission from M. B. Hall, *Inorg. Chem.* **17** 2264 (1978). Copyright by the American Chemical Society.

case if the energy changes of the $1b_2$ orbital on bending (involved purely in Ap–ligand σ interactions) is small (Figure 7.13), then two valence pair molecules may be nonlinear. For SrF_2 and BaF_2 where the $ns–np$ separation is small, bent molecules are found experimentally. This is in opposition to the VSEPR result which would predict a linear structure (AB_2) and is also not in accord with the result suggested by viewing the molecules on an ionic model as $F^- A^{2+} F^-$ (see the argument for Li_2O in Section 3.1).

Also of interest is that the planar or pyramidal structure of NH_3 and PH_3 is also very susceptible to the $ns–np$ separation (Figure 7.15). Pyramidal geometries are found for larger values of this parameter. These AH_3 molecules were predicted to be pyramidal in Section 6.2 by consideration of the quartic (f_σ) energy term on the AOM model with p orbitals alone. Inclusion of s-orbital interaction with the $2a_1$ orbital (correlating with the $1a_2''$ orbital of the planar structure) disfavored the pyramidal structure by the switching on of an As–ligand σ antibonding contribution (Figure 6.6b). As the ns–np separation decreases and/or the interaction between the two centers gets stronger, this antibonding effect will result in a decrease of the barrier to inversion until eventually the planar structure is more stable. For BH$_3$, where the HOMO ($1e'$) does not contain any As orbital character at either planar or pyramidal geometries, the geometry (planar) is independent (Figure 7.15) of the ns–np separation (as long as it is positive).

For SH$_4$ the calculated geometry is extremely sensitive to the $3p$–$3s$ energy separation and the C_{2v} and C_{4v} structures are very close in energy. Here the extent of s-orbital participation in the HOMO closely controlled its shape in Section 5.4. For ArH$_2$ (XeF$_2$), the As orbital is calculated to be important in a different way. If the energy of the $2\sigma_g^+$ orbital is higher than that of $2\sigma_u^+$, then the HOMO in ArH$_2$ will be stabilized on bending (Figure 4.2). Recall that in Chapter 4 we rationalized the linear XeF$_2$ geometry on the basis that $2\sigma_g^+$ was the HOMO in this system. Thus a bent XeF$_2$ species is predicted for a small np–ns separation due not to a change in a_1 orbital curvature but by a change in configuration. A similar effect is predicted to occur with ArH$_4$ (XeF$_4$). In general then inclusion of the central-atom s orbital controls the molecular geometry in very much the way we described in Chapter 6 and the np–ns separation (as an indicator of the relative importance of s-orbital participation) often influences the geometry by controlling the curvature of a_1 levels on distortion. EHMO calculations performed with[74] and without [67] the inclusion of overlap integrals between the nonbonded atoms give qualitatively very similar geometric results. This suggests that usually we are not making too drastic an approximation in neglecting them in our simple models.

An interesting series of results is found in AH$_n$ molecules when d orbitals are included on the central atom. These *may* be of importance only for second-row and heavier atoms. For the high-symmetry geometries ArH$_2$, SiH$_4$, ArH$_4$, PH$_5$, and SH$_6$, as the $3d$ orbitals become more important the barriers to distortion decrease. But for the molecules ClH$_3$, SH$_4$, ClH$_5$, and ArH$_6$, the addition of the d orbitals tended to adjust the geometry so that it was closer to the one where all HAH angles were either 90° or 180°. Similar effects are observed in calculations using either EHMO or *ab initio* methods. Thus the d orbitals appear not to be critical for gross geometry determination. Recall that in the directed valence approach of Section 3.2, d orbitals were mixed on an equal basis with s and p orbitals to form hybrids when more than four pairs were present. These higher-energy orbitals are however probably of importance in the fine tuning of the geometry and in determining the barriers to geometric distortions. Inclusion of d orbitals on the central atom will result in an increase in the amount of electron density residing there at the expense of that on the coordinated atoms. If central-atom d orbital participation were extensive, then we might expect to see the most electronegative atom as the central moiety in the structure. Experimentally there is spectroscopic evidence for d-orbital participation in the ground-state wavefunction (X-ray spectroscopy, esr, etc.), but how important it is energetically is difficult to conclude from these studies.

8 Theories of Transition Metal–Ligand Interaction: The Crystal Field Model

One of the earliest approaches for viewing transition metal–ligand interaction that provides a useful way of understanding electronic spectral and thermodynamic properties is an ionic model, the crystal field theory (CFT).[7,8,18,34,53,70,80,156,190,205,206] This scheme places n point charges at the ligand positions, and the resultant electrostatic interaction between these charges and the metal d electrons may be used after some manipulation to view the intensity and frequency of the "d–d" electronic absorption spectra of the MY_n species. This ionic model is perhaps an unrealistic way of representing MY interaction but has performed well in the spectral field. As we shall see the scheme is not a good way to view structural chemistry. An improved version, the adjusted crystal field theory (ACFT), included several extra terms to make the approach more "realistic" but fell short of actually including covalent interaction. The term "Ligand Field Theory" (LFT) is generally used to rationalize phenomena in which the ligands exert some effect that cannot be regarded as being due to the specific interaction of a ligand charge or negative end of a dipole with the d electrons. σ and π bonding for example have no place in such an ionic model, and it is lack of inclusion of these terms that makes the CFT or its variants chemically unacceptable and led to the use of the LFT—really molecular orbital theory applied to transition metal systems.[8,44,207] There are at least five[53,116] experimental reasons which are often quoted that emphasize the importance of including these covalent terms.

1. Analysis of the optical spectra of inorganic complexes shows that the Racah parameters (containing combinations of the interelectronic repulsion integrals) are lower than the free ion values.

2. EPR spectra show that the spin–orbit coupling constant of the metal is less in the complex than in the free ion.

3. The orbital contribution to the magnetic moment is smaller in the complex than in the free ion.

4. In $IrCl_6^{2-}$ hyperfine coupling is observed between the odd electron and the chlorine nucleus. (Similar behavior is found in other systems, e.g., hyperfine coupling to ^{13}C in $Mo(^{13}CN)_8^{3-}$.)

5. The ^{19}F resonance frequency in MnF_2 is very different from that of free F^-.

In the following chapters we develop molecular orbital ideas which naturally include covalent bonding to describe MY interactions and compare the results obtained in the structural and kinetic fields with those achieved using the CFT.

8.1 The CFT in Cubic Fields

There are many in-depth treatments of this approach, and our strategy here is to give a rather cursory survey of the method but emphasizing important and sometimes little discussed aspects of the model. Some writers[109] have dismissed the Crystal Field Theory as being an incorrect view of transition metal–ligand interaction that should not be used at all. It does provide a parametrized way to view transition metal chemistry and, as we shall see later, has some strong resemblances to the Angular Overlap Approach. We include it here since the derivation of the form of the d-orbital splitting energies and the application of the results are of value in themselves. The basic principles of the CFT were devised by Bethe and Van Vleck.[18,205,206] They considered the effect of the electrostatic potential on the energy levels of a free ion when it is located in a crystal lattice and is surrounded by an electric field of a given symmetry. By means of group theory they showed how degenerate orbitals and atomic terms split apart in energy in a way dependent on the ion's site symmetry. Later the idea of replacing the lattice by a collection of nearest neighbor points, namely, ligands, led to the development of the CFT in molecular coordination chemistry.

The qualitative effects of the crystal field are relatively easy to see. Thus in Figure 8.1 for the octahedral MY_6 system the one-electron d-

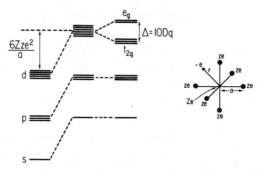

Figure 8.1 Effect of an octahedral crystal field on the atomic s, p, and d orbitals of the metal. All orbitals are raised equally in energy by a spherical field, but the d orbitals are split apart in energy as the symmetry is lowered to octahedral.

orbital energies are firstly all raised a uniform amount by the presence of the six ligand charges and then split into two sets—those to lower energy where the lobes point between the ligands (t_{2g} set: xy, xz, yz*) and receive less electrostatic repulsion from the ligand charges, and those where the lobes point at the ligands (e_g set: x^2-y^2, z^2) and receive a large electrostatic repulsion (Figure 8.2). Removal of one ligand from the octahedron removes the e_g degeneracy ($a_1 + b_1$) and partially lifts the degeneracy of the t_{2g} set ($e + b_2$) as in Figure 8.2. The electron in the z^2 orbital receives less destabilization relative to x^2-y^2 since one of the charges pointing directly at one of the lobes along the z-axis has been removed. What is less easy to see in qualitative terms is why x^2-y^2 experiences a larger CF destabilization in the square pyramid compared to the octahedron. This is connected with the fact that the barycenter, or center of gravity, of the crystal field energy levels remains invariant to the coordination geometry, number, type of ligands, and so on. In order

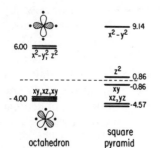

Figure 8.2 Crystal field splitting of metal d orbitals in octahedral and square pyramidal ($Y_{ap}MY_{ba}$ angle $= 90°$) geometries (units of Dq; values for spy are for $\rho = 2$, see p. 118), showing for the octahedral structure how the ligand charges lie between the lobes of xz, yz, and xy but point at the lobes of x^2-y^2 (and z^2).

* We shall write d_{z^2}, d_{xy}, etc. orbitals as z^2, xy, etc.

to see how this arises we need to calculate quantitatively the relevant splitting energies, as we show later.

The $e_g - t_{2g}$ energy separation, labeled $10Dq$ or Δ, may be readily calculated from electronic spectral measurements.[53,61,108-110,130,179] The electronic energy of the ground state is related to that of an excited state by the sum of two terms. The first is the difference in orbital energy between the ground- and excited-state electron configurations (measured as some multiple or fraction of Δ), and the second is the difference between the interelectronic repulsion forces in the two states, usually represented in terms of the Racah parameters. By analysis of electronic spectra these parameters may be obtained numerically for a given system. (The exact dependence of state energy on the "orbital" (one electron) and electronic repulsion (two electron) terms is more complex than we describe here, but this partitioning will serve our purpose). A very interesting multiplicative relationship correlates the results obtained for a wide variety of systems. Jørgensen[105,107] has showed that Equation 8.1 holds quite well:

$$f \times g \times 10^3 = \Delta \text{ cm}^{-1} \qquad (8.1)$$

f is a value characteristic of the central metal atom and its formal oxidation state and g is a value characteristic of the ligand. Table 8.1 shows some f and g values for some octahedral MY_6 and chelated $M(Y-Y)_3$ systems.

Evaluation of the crystal field splitting patterns in a quantitative manner is a process of some mathematical complexity, more tedious however than difficult.[70,80,190] Firstly we need to set up a suitable expression for the total crystal field V which reflects the geometric arrangement of the ligands. This is usually done as a series expansion in terms of spherical harmonics. Perturbation theory is then used to evaluate the integrals or matrix elements $\int \Psi_i V \Psi_j \, d\tau$, where the Ψ_i are the wavefunctions describing the n electron system. Since we are interested here in orbital energies our discussion will emphasize the one-electron case where $\Psi_i = \psi_{d_i}$, namely a d^1 system. Solution of a 5×5 secular determinant of the type of Equation 1.16 gives five values of $E-E_d$ which are the shifts in the one-electron d-orbital energies. As an example we take the octahedral system and show the procedure in a very condensed fashion.

First we assume some unspecified electrostatic potential between the ligands and the central atom. From group theory the form of the cubic potential is given by

$$V = f(r)\left[Y_{40} + \left(\frac{5}{14}\right)^{1/2} (Y_{44} + Y_{4-4}) \right] \qquad (8.2)$$

Table 8.1 Jørgensen's f and g Values for Some Ligands and Central Metal Atoms

Ligand	f Factor	Electronic Configuration	Metal	g Factor
Br^-	0.72	$3d^5$	Mn(II)	8.0
SCN^-	0.73	$3d^8$	Ni(II)	8.7
Cl^-	0.78	$3d^7$	Co(II)	9
N_3^-	0.83	$3d^3$	V(II)	12.0
F^-	0.9	$3d^5$	Fe(III)	14.0
DMSO [$(CH_3)_2SO$]	0.91	$3d^3$	Cr(III)	17.4
Urea [$(NH_2)_2CO$]	0.92	$3d^6$	Co(III)	18.2
CH_3COOH	0.94	$4d^6$	Ru(II)	20
C_2H_5OH	0.97	$3d^3$	Mn(IV)	23
H_2O	1.00	$4d^3$	Mo(III)	24.6
NCS^-	1.02	$4d^6$	Rh(III)	27.0
$p\text{-}CH_3C_6H_4NH_2$	1.15	$4d^3$	Tc(IV)	30
NC^-	1.15	$5d^6$	Ir(III)	32
CH_3NH_2	1.17	$5d^6$	Pt(IV)	36
CH_3CN	1.22			
C_5H_5N	1.23			
NH_3	1.25			
En	1.28			
Dien	1.29			
NH_2OH	1.30			
Dipy	1.33			
Phen	1.34			
CN^-	~1.7			

Adapted from Refs. 108 and 110.

This applies equally well to octahedral, tetrahedral, and cubal (MY_8) coordination until we specify the form of $f(r)$, that is, where the charges lie and the exact nature of the potential. Y_{lm} are the spherical harmonics and contain terms in θ, ϕ only as a simple product $g(\theta)$ $h(\phi)$. Recall that the form of the d orbitals contains a similar division (Section 1.2)

$$\psi_d = R(r)\Theta(\theta)\Phi(\phi) \tag{8.3}$$

where Θ, Φ are easily written down but the radial dependence $R(r)$ varies (a) from atom to atom and (b) depends on the atomic charge and (c) on the details of the coordination geometry. This means that each matrix element $\int \psi_{d_i} V \psi_{d_j} \, d\tau$ may be written as the product of three integrals each over one variable (Equation 8.4):

$$\int f(r)R^2(r)\ r^2 dr \int g(\theta)\ \Theta^2(\theta)\ \sin\theta\ d\theta \int h(\phi)\Phi^2(\phi)\ d\phi \qquad (8.4)$$

The second and third terms in this expression are readily evaluated—simple integrals containing θ and ϕ are found. (In practice the easiest way of evaluating these integrals is to make use of some identities concerning products of the spherical harmonics.)

The matrix elements, ready for insertion into the secular determinant, are found to be

$$\int \psi_{d_0} V \psi_{d_0}\ d\tau = 6y$$

$$\int \psi_{d_{\pm 1}} V \psi_{d_{\pm 1}}\ d\tau = -4y$$

$$\int \psi_{d_{\pm 2}} V \psi_{d_{\pm 2}}\ d\tau = y \qquad (8.5)$$

$$\int \psi_{d_{\pm 2}} V \psi_{d_{\mp 2}}\ d\tau = 5y$$

All the other elements (e.g., $\int \psi_{d_0} V \psi_{d_{\pm 1}}\ d\tau$) are equal to zero; y is a term which contains the integral $\int f(r)\ R^2(r)\ r^2\ dr$ and a constant which has arisen via the θ, ϕ integrations. Solution of the secular determinant leads to the energy levels $E = +6y$ (twice) and $E = -4y$ (three times) shown schematically in Figure 8.3 and is the familiar CF result for both octahedral and tetrahedral geometries (y of course has a different sign in the two systems). However recall that as yet we have not defined the form of the electrostatic potential. The splitting of the levels in this figure is purely a result of symmetry arguments. What we find from character tables ($d \rightarrow t_{2g} + e_g$ or $t_2 + e$) we have worked out longhand by the specification of V in Equation 8.2. Evaluation of y requires a description of the exact nature of the electrostatic potential and actually placing the charges at points in space (i.e., the $f(r)$ in Equation 8.2). The simplest model that can be used is to represent the ligands as point negative charges (ze) situated at a distance a from the central ion (charge $+Ze$) and evaluation of the integral $\int f(r)\ R^2(r)\ r^2\ dr$. Since we do not know the function $R(r)$, this is left as a parameter. For the octahedral case we find $y = \frac{1}{6}(Zze^2/a^5)\langle r^4 \rangle$; for the tetrahedral system, $y = \frac{4}{9} \cdot \frac{1}{6}(Zze^2/a^5)\langle r^4 \rangle$; and

e_g, e

$6y$

$4y$

t_{2g}, t_2

Figure 8.3 d-Orbital splitting in cubic fields; $y = Dq$ for the MY_6 octahedron; $y = -4/9 Dq$ for the MY_4 tetrahedron; and $y = -8/9 Dq$ for the MY_8 cube.

for the cube, twice this value. (This is easy to see; the cube is simply two intermeshed tetrahedra.) $\langle r^4 \rangle$ is the mean fourth power radius,

$$\langle r^4 \rangle = \int_0^\infty R(r)\, r^4 R(r)\, r^2\, dr \tag{8.6}$$

and contains the one unknown of the model. The quantity y_{oct} is referred to as Dq which leads to an e_g–t_{2g} splitting of $10Dq$ in the octahedron, $-\frac{8}{9} \cdot 10Dq$ in the cube, and e–t_2 splitting of $\frac{4}{9} \cdot 10Dq$ in the tetrahedron. The factor of $\frac{8}{9}$ or $\frac{4}{9}$ appearing here is known as the geometric factor since it arises purely as a result of the number and angular disposition of the ligands and is independent of the nature of the electrostatic potential we choose.

Dq has been estimated by evaluation of Equation 8.6 using available $3d$ metal wavefunctions. Van Vleck's initial quantitative comparisons between observed and calculated values of Dq for $Cr(H_2O)_6^{3+}$ were in favorable agreement. However Kleiner[120] used a more "realistic" approach where the ionic model was retained but the ligands regarded as having some spatial extent. This led to contributions to Dq of ~1000 cm^{-1} from the point dipoles and ~ – 1500 cm^{-1} from "Kleiner's correction," namely the smearing out of the electronic charge. This calculated value of Dq of – 500 cm^{-1} is of the opposite sign to that observed (1750 cm^{-1}). Repeating the calculation but with a better Cr $3d$ wavefunction leads to a positive but small value for Dq, a long way from experiment.[65] Assuming that the H_2O molecules were point dipoles, and neglecting Kleiner's correction, the observed value of Dq can only be matched if an unrealistic dipole moment many times that found experimentally is used. Thus the quantitative basis of the approach is not very good. Considerably better agreement with experiment has been found using molecular orbital (i.e., covalent) methods.

8.2 Lower-Symmetry Fields

In all other geometries the noncubic nature of the potential means that other spherical harmonics need to be introduced to define V. The overall result is the introduction of an extra term, one containing $\langle r^2 \rangle$ into the general expressions describing the d-orbital energies. There are a variety of ways this is done.[70] In addition to the definition

$$Dq = \frac{1}{6} \frac{Zze^2}{a^5} \langle r^4 \rangle \tag{8.7}$$

we may define

$$Cp = \frac{2}{7} \frac{Zze^2}{a^3} \langle r^2 \rangle \tag{8.8}$$

For the squashed tetrahedron of D_{2d} symmetry (bond angles of 2β) for example, the relevant matrix elements are

$$\int \psi_{d_0} V \psi_{d_0} \, d\tau = \frac{6}{7} X + 2Y$$

$$\int \psi_{d_{\pm 1}} V \psi_{d_{\pm 1}} \, d\tau = -\frac{4}{7} X + Y \tag{8.9}$$

$$\int \psi_{d_{\pm 2}} V \psi_{d_{\pm 2}} \, d\tau = \frac{1}{7} X - 2Y$$

$$\int \psi_{d_{\pm 2}} V \psi_{d_{\mp 2}} \, d\tau = -Z$$

where $X = (35 \cos^4 \beta - 30 \cos^2 \beta + 3)Dq$, $Y = (3 \cos^2 \beta - 1)Cp$, and $Z = 5Dq \sin \beta$. Obviously the terms in Cp disappear at $\beta = \cos^{-1}(1/\sqrt{3})$, the tetrahedral geometry, where only Dq is needed to define the energies. In C_{4v} distortions of octahedral complexes, either for the case where the axial and equatorial MY distances are different or where the ligands are chemically different, the parameters Ds and Dt are useful ones to use. These describe the size of the distortions from octahedral symmetry:

$$Ds = Cp(a) - Cp(b) \tag{8.10}$$

$$Dt = \frac{4}{7} [Dq(a) - Dq(b)]$$

The a and b labels refer to axial and equatorial ligands. For the Cp parameters we define

$$Cp(a) = \frac{2}{7} \frac{Zze^2}{a^3} \langle r^2 \rangle \tag{8.11}$$

$$Cp(b) = \frac{2}{7} \frac{Zz'e^2}{b^3} \langle r^2 \rangle$$

where we have distinguished the two different sorts of ligands by using different MY distances a, b and different ligand charges z, z' (not necessarily integral).

Analysis of the electronic spectra of these noncubic molecules leads to determination of values of the Dq's, Cp's, Ds, and Dt but usually after much more effort than in the octahedral and tetrahedral case. Table 8.2 shows experimental values for these parameters for three D_{2d} molecules. Quite a difference in the ratio of Cp/Dq is found between the copper and nickel systems. We should note however in this context that a simple cancellation of the form of Equation 8.12 is *not* valid:

$$\frac{Cp}{Dq} = \frac{12}{7} \frac{a^2}{\langle r^2 \rangle} \sim \frac{12}{7} = 1.71 \qquad (8.12)$$

In general then the d-orbital energies are defined by two parameters (for a binary MY_n system), but only one is needed in the cubic environment. Cp does appear generally to be larger than Dq, and there seems to be, from an analysis of rather a limited set of data, an inverse correlation between Dq and Cp/Dq for a given series of ligands, that is, as Dq increases Cp decreases. For example[70] from the analaysis of a series of molecules FeN_4Y_2, the order in Dq was $Y = py > Cl > Br$ for $N = py$ and $Y = $ isoquinoline $> Cl > Br > I$ for $N = $ isoquinoline. For Cp/Dq the order was the reverse.

Table 8.2 Crystal Field Parameters [a] From Electronic Spectral Data (Units 1000 cm^{-1})

D_{4h} $MY_4Y'_2$ Complex	Equatorial Ligands Y			Axial Ligands Y'		
	Dq	Cp	Cp/Dq	Dq	Cp	Cp/Dq
Nipyridine$_4$Cl$_2$	1.173	1.497	1.28	0.678	1.006	1.48
Nipyrazole$_4$Cl$_2$	1.096	1.957	1.79	0.610	0.834	1.37
Nipyridine$_4$Br$_2$	1.150	1.429	1.24	0.599	0.797	1.33
Nipyrazole$_4$Br$_2$	1.092	1.940	1.78	0.498	0.634	1.27
Cr(en)$_2$F$_2^+$	2.170	2.067	0.95	1.610	2.867	1.78
Cr(en)$_2$(H$_2$O)$_2^{3+}$	2.350	2.238	0.95	1.685	2.545	1.51
Cr(en)$_2$Cl$_2^+$	2.250	2.143	0.95	1.341	1.971	1.47
Cr(en)$_2$Br$_2^+$	2.250	2.143	0.95	1.236	1.677	1.37
D_{2d} MY_4 complex						
CuCl$_4^{2-}$	1.297	2.456	1.89			
CuBr$_4^{2-}$	1.056	1.553	1.47			
NiCl$_4^{2-}$	0.734	4.790	6.53[b]			

[a] Adapted from Ref. 70.

[b] For comment on the accuracy of these results, see Ref. 70, p. 83.

8.3 The Meaning of the CF Parameters

It is convenient to list the molecular parameters of the CFT that may differ from one another in real systems. First we acknowledge[70] that some covalent bonding has occurred to alter the charge distribution in the molecule from the purely ionic model (i.e., Z, z, z' may be nonintegral), but we still view the origin of d-orbital splittings by using the CF model.

1. Cp and Dq (Equations 8.7 and 8.8) involve the averages $\langle r^2 \rangle$ and $\langle r^4 \rangle$, respectively, integrals involving the radial part of the d-orbital wavefunction $R(r)$. Whatever its precise form, we know that increasing the metal charge will lead to orbital contraction and decreasing the charge will lead to just the reverse. Because of the difference in exponentiation in $\langle r^n \rangle$ we expect $\langle r^4 \rangle$ to be more sensitive to such changes and hence a larger decrease in Dq than in Cp on increasing the metal charge Z. Thus an increase in metal charge Z leads to a decrease in Dq and an increase in Cp/Dq.

2. Decreasing the ligand charge z will lead to the same proportional decreases in Dq and Cp. (i.e. decrease in Dq, Cp/Dq unchanged).

3. Increasing the bond length will lead to a larger decrease in Dq than in Cp (i.e., Dq decreases, Cp/Dq increases).

4. Larger values of Dq and Cp should be found for larger atoms with a proportionately larger effect for Dq.

5. Increasing the metal ionization potential, often associated with a decrease in size (d), should lead to a decrease in Dq and Cp.

The general trend in Cp and Cp/Dq mentioned in the previous section is in general agreement with these predictions (i.e., as Dq increases, Cp/Dq decreases) with the exception of (2) which left Cp/Dq unchanged. There are exceptions to this rule, for example, the copper halides of Table 8.2. Crystallographic data show a longer CuBr bond than the CuCl bond. Literal interpretation of the crystal field parameters, (3) above, implies a smaller $Dq(Br)$ than $Dq(Cl)$ which is found, but also a Cp/Dq ratio which changes in the opposite sense to that found in Table 8.2. The bromo complex is perhaps more covalently bound (in molecular orbital language this means a better σ donor) than the chloro analog. Thus the effective nuclear charge should be less in $CuBr_4^{2-}$ than in $CuCl_4^{-2}$. This leads to a larger $Dq(Br)$ than $Dq(Cl)$ on prediction (1) and also to a decreased Cp/Dq for the bromo compared to chloro complex, which is the opposite result to that just predicted on factor (3). Since the Br ligand will carry less charge than the Cl ligand, from (2) $Dq(Br) < Dq(Cl)$. To summarize:

1. $Dq(Br) > Dq(Cl)$ $Cp/Dq(Br) < Cp/Dq(a)$
2. $Dq(Br) < Dq(Cl)$ $Cp/Dq(Br) = Cp/Dq(Cl)$
3. $Dq(Br) < Dq(Cl)$ $Cp/Dq(Br) < Cp/Dq(Cl)$

In the present circumstances we *could* rationalize the copper halide data on the grounds that (1) is more important than (3) such that in the two opposed trends $Dq(Br) > Dq(Cl)$ and $Cp/Dq(Br) < Cp/Dq(Cl)$. Inclusion of factor (2) can reverse the trends in the Dq values but leave the Cp/Dq ratio trend unchanged. For the more general inverse dependence of Dq and Cp/Dq we assume that factor (2) is smaller than the result of the opposed trends (1) and (3).

The variation in Dq across the first transition series is irregular as may be seen from Jørgensen's table (8.1). From effect (5) we should expect f to decrease across the series from left to right, although of course there are bond length changes and charges to take into account as well. The increase in Dq on going down a group as the size of the atom increases, effect (4), is well documented but Dq usually increases with formal charge whereas the theory predicts the reverse although the effects of increasing Z and contracting the wavefunction work in opposite directions.

Most experimental studies have centered on octahedral and tetrahedral molecules, and at the present time there is not a complete set of Dq and Cp data for a very wide range of heavier systems to properly assess all of these predictions. There are clearly objections to the point charge ionic model on chemical grounds, and this analysis has tried to remove them by including some simple ideas which are molecular orbital based. However there are enough adjustable parameters in the model and five different (1–5) effects which may be included so that almost any set of data may be "rationalized" as in our discussion of the copper halides. The chemical worth of such procedures as to how they improve our understanding of these systems needs yet to be established. The CFT however is still a useful method to view and catalog spectral features via the splitting of atomic term energies by ligand coordination.

8.4 The Derivation of Orbital Diagrams for Noncubic Geometries

The mathematical expression for the potential field at the central metal atom for a MY_n complex of a given geometry may be simply obtained by summing the potentials due to two or more smaller units MY_{n_1} and MY_{n_2} where $n_1 + n_2 \equiv n$ and the arrangements in space of the two ligand sets are additive as in the example of Figure 8.4. We have seen an example

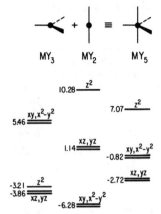

Figure 8.4 Illustration of d-orbital energy additivity rule. The d-orbital energies (units of Dq) of the trigonal bipyramid are each equal to the sum of the corresponding values in the trigonal planar MY_3 and linear MY_2 molecules (values for $\rho = 2$).

of this type in Chapter 8-1 where the $t_{2g}-e_g$ separation in the cube (two intermeshed tetrahedra) was exactly twice as large as the t_2-e separation in the tetrahedron. Thus if the crystal fields of a small number of primary geometries are known then the potentials experienced by larger units may be obtained by simple addition. Two rules control the application of the method devised by Krishnamurthy and Schaap.[124,225]

Rule 1. The d-orbital levels of the CFT for MY_n may be readily derived by simple addition of the d-orbital energies of the constituent units MY_{n_1} and MY_{n_2}. As it turns out for a large number of systems, the results for three primary units only are needed: those of the MY, MY_2 (with $YMY = 90°$), and $D_{2d}MY_4$ geometries. The d-orbital energies for these geometries are given in Table 8.3, the last entry as a function of geometry. Two sets of values of the d-orbital energies are given in terms of the parameter ρ, which relates the $\langle r_n \rangle$ parameters (Equation 8.13) :

$$\rho = a^2 \frac{\langle r^2 \rangle}{\langle r^4 \rangle} = \frac{7}{12} \frac{Cp}{Dq} \tag{8.13}$$

For $\rho = 1$ and 2, the ratio of Cp/Dq is 1.71 and 3.43, respectively. Two examples will show the application of the method. For the square planar molecule we use the addition process of Figure 8.5.

$\rho = 2$	z^2	x^2-y^2	xy	xz	yz
A	−2.14	6.14	1.14	−2.57	−2.57
B	−2.14	6.14	1.14	−2.57	−2.57
Sum ≡ square plane	−4.28	12.28	2.28	−5.14	−5.14

Table 8.3   Crystal Field d Orbital Energies for the Three Primary
Groups (Units of Dq)
A. Groups I and II

	Relative Energy of d Orbitals for $\rho = 2.0$ (Values for $\rho = 1.0$ in Parentheses)				
	z^2	x^2-y^2	xy	xz	yz
I. MY (C_{2v}) One ligand or charge located on the z-axis	5.14 (3.43)	-3.14 (-1.43)	-3.14 (-1.43)	0.57 (-0.285)	0.57 (-0.285)
II. MY$_2$ (C_{2v}) Two ligands at right angles in the $x-y$ plane (on the x- and y-axes)	-2.14 (-0.43)	6.14 (4.43)	1.14 (-0.57)	-2.57 (-1.715)	-2.57 (-1.715)

B. Group III

	Orbital	Energy
III. MY$_4$ (D_{2d}) Four ligands in a staggered, tetrahedral-like arrangement, two above and two below the $x-y$ plane; each at a variable angle β with respect to the $\pm z$-axes	z^2	$\frac{9}{8}[35 \cos^4 \beta + 6 \cos^2 \beta(2\rho - 5) - 4\rho + 3]$
	x^2-y^2	$\frac{3}{4}[35 \cos^4 \beta - 6 \cos^2 \beta(12\rho + 5)$ $\mp 35 \sin^4 \beta + 24\rho + 3]$
	xy	$\frac{3}{4}[35 \cos^4 \beta - 6 \cos^2 \beta(12\rho + 5)$ $\pm 35 \sin^4 \beta + 24\rho + 3]$
	xz	$\frac{3}{4}[35 \cos^4 \beta - 3$ $\cos^2 \beta(3\rho + 10\rho) + 3\rho + 3]$
	yz	$\frac{3}{4}[35 \cos^4 \beta - 3$ $\cos^2 \beta(3\rho + 10) + 3\rho + 3]$

For the octahedron we need to perform the summation of Figure 8.6.

$\rho = 2$	z^2	x^2-y^2	xy	xz	yz
A	5.14	-3.14	-3.14	0.57	0.57
B	5.14	-3.14	-3.14	0.57	0.57
C	-2.14	6.14	1.14	-2.57	-2.57
D	-2.14	6.14	1.14	-2.57	-2.57
Sum = octahedron	6.00	6.00	-4.00	-4.00	-4.00

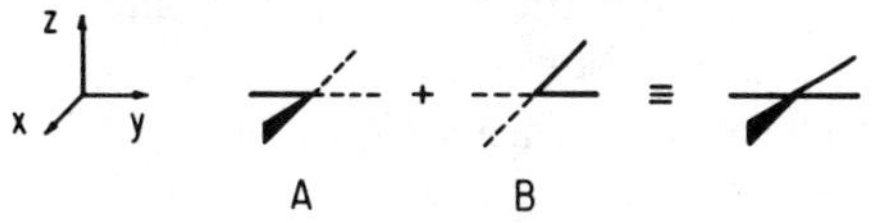

Figure 8.5 Derivation of square planar MY_4 energy levels from two cis MY_2 species.

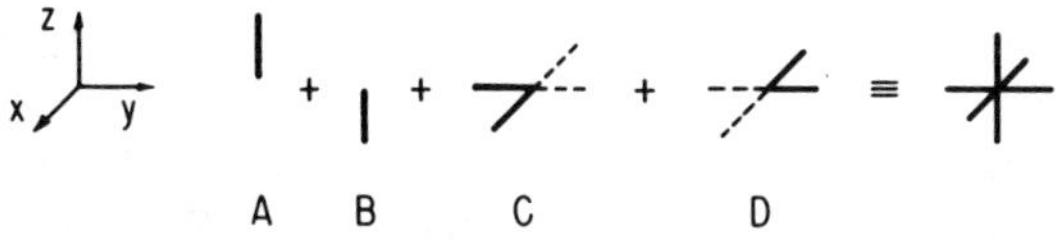

Figure 8.6 Derivation of octahedral MY_6 energy levels.

This particular result of course is independent of the value of ρ since the cubic system needs only one parameter (Dq) to describe the d-orbital energies. For the five-coordinate square pyramid (axial–basal angle = 90°) we may just subtract the contribution of one axial ligand from the octahedral values (Figure 8.7).

$\rho = 2$	z^2	x^2-y^2	xy	xz	yz
Octahedron	6.00	6.00	−4.00	−4.00	−4.00
A	5.14	−3.14	−3.14	0.57	0.57
Difference = square pyramid	0.86	9.14	−0.86	−4.57	−4.57

It is now possible to see the origin of the extra destabilization afforded x^2-y^2 in the square pyramid compared to the octahedron mentioned earlier. The extension of the scheme to other geometries is usually straightforward. We must however make sure that the ligands are arranged correctly with respect to the axes chosen in Table 8.3 for a particular geometry. Thus in the case of the square plane all four ligands must lie in the xy plane, and for the square pyramid the unique ligand should be located along the z-axis to get the conventional labelling of the d orbitals. The method is however inapplicable to low-symmetry situations where two d orbitals may transform as the same representation. For example in the T-shaped molecule (C_{2v}), both z^2 and x^2-y^2 transform as a_1 (z-axis parallel to the upright of the T) and will mix together. Two new orbitals

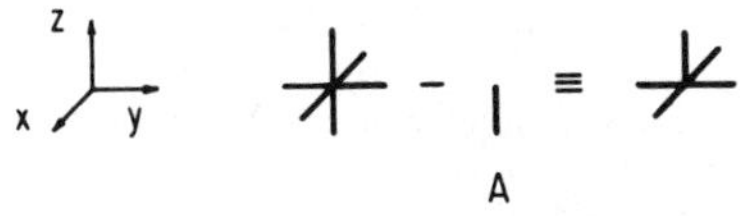

Figure 8.7 Derivation of square pyramidal MY_5 energy levels.

will be produced; mixtures of z^2 and x^2-y^2 and their energies in this potential field are not readily obtained on this scheme.

For some geometries we need to make use of a second rule.

Rule 2. The *total* electrostatic effect of any array of ions (ligands) lying along a particular axis or in a given plane on all the orbitals of the central atom is independent of the geometric arrangement of ions in the array. We may use this fact, combined with a knowledge of the qualitative way the orbitals split apart in energy in a given symmetry field, to calculate orbital energies in several other geometric types. For example consider the (unknown) hexagonal MY_6 coordination D_{6h}. From inspection of a set of character tables we know that the d orbitals split into a_{1g} (z^2), e_{2g} (xy, x^2-y^2) in plane orbitals, and e_{1g} (xz, yz) out-of-plane orbitals. (Table 8.4 gives a set of symmetry species for central-atom s, p, and d orbitals for a variety of geometries.) Thus in order to evaluate the effect of these orbitals we model the hexagon as the sum of three pairs of ligands at right angles (Figure 8.8) (there are of course alternative spatial arrangements to the ones we show) and average the energy contributions to xy, x^2-y^2 which

$\rho = 2$	z^2	x^2-y^2	xy	xz	yz
A	-2.14	6.14	1.14	-2.57	-2.57
$\times\ 3\ =\ B$	-6.42	18.42	3.42	-7.71	-7.71
Average	-6.42	10.92	10.92	-7.71	-7.71

we know are degenerate. The six charges located in the plane have the same effect on orbitals not located in the plane (z^2, xz, yz) irrespective of their arrangement. The *total* crystal field interaction with these in-plane orbitals is independent of orientation but the individual interactions depend upon the orientation of the charges. Under D_{6h} symmetry they are degenerate and we simply average the individual contributions on the 3 $\times$ A model.

The above result allows a simple derivation of the levels of the trigonal plane. Because the degeneracies of the d-orbital levels are similar in D_{6h} and D_{3h} the crystal field energies for trigonal planar MY_3 are simply one half of the values for the hexagonal plane. Table 8.5 gives some d-orbital

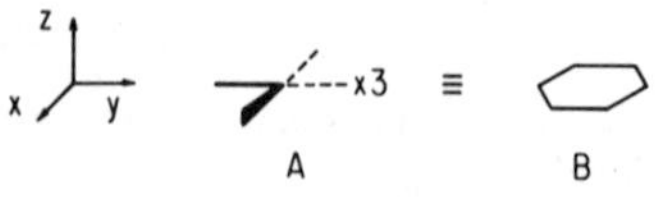

Figure 8.8 Derivation of hexagonal MY_6 energy levels.

Table 8.4 Transformation Properties of Metal s, p and d Orbitals For Various Geometries

Geometry	s	p_x	p_y	p_z	z^2	x^2-y^2	xy	xz	yz
MY linear $C_{\infty v}$	a_1	b_1	b_2	a_1	a_1	a_1	a_2	b_1	b_2
MY$_2$ linear $D_{\infty h}$	σ_g^+	π_u	———	σ_u^+	σ_g^+	δ_g	———	π_g	———
Bent C_{2v}	a_1	b_1	b_2	a_1	a_1	a_1	a_2	b_1	b_2
MY$_3$ fac trivacant C_{3v}	a_1	e	———	a_1	a_1	e	———	e	———
Trigonal plane D_{3h}	a_1'	e'	———	a_2''	a_1'	e'	———	e''	———
T-shape C_{2v}	a_1	b_1	b_2	a_1	a_1	a_1	a_2	b_1	b_2
MY$_4$ tetrahedron T_d	a_1	t_2	———	———	e	———	t_2	———	———
Square plane D_{4h}	a_{1g}	e_u	———	a_{2u}	a_{1g}	b_{1g}	b_{2g}	e_g	———
Trigonal pyramid, C_{3v}	a_1	e	———	a_1	a_1	e	———	e	———
Cis divacant, C_{2v}	a_1	b_1	b_2	a_1	a_1	a_1	a_2	b_1	b_2
MY$_5$ pentagonal plane D_{5h}	a_1'	e_1'	———	a_2''	a_1'	e_2'	———	e_1''	———
Trigonal bipyramid D_{3h}	a_1'	e'	———	a_2''	a_1'	e'	———	e''	———
Square pyramid C_{4v}	a_1	e	———	a_1	a_1	b_1	b_2	e	———
MY$_6$ octahedron, O_h	a_{1g}	t_{1u}	———	———	e_g	———	t_{2g}	———	———
Hexagonal plane, D_{6h}	a_{1g}	e_{1u}	———	a_{2u}	a_{1g}	e_{2g}	———	e_{1g}	———
Pentagonal pyramid, C_{5v}	a_1	e_1	———	a_1	a_1	e_2	———	e_1	———
MY$_7$ heptagonal plane, D_{7h}	a_1'	e_1'	———	a_2''	a_1'	e_2'	———	e_1''	———
Pentagonal bipyramid, D_{5h}	a_1'	e_1'	———	a_2''	a_1'	e_2'	———	e_1''	———
Hexagonal pyramid, C_{6v}	a_1	e_1	———	a_1	a_1	e_2	———	e_1	———
MY$_8$ cube, O_h	a_{1g}	t_{1u}	———	———	e_g	———	t_{2g}	———	———
Square antiprism D_{4d}	a_1	e_1	———	b_2	a_1	e_2	———	e_3	———
Octagonal plane D_{8h}	a_{1g}	e_{1u}	———	a_{2u}	a_{1g}	e_{2g}	———	e_{1g}	———
Hexagonal bipyramid, D_{6h}	a_{1g}	e_{1u}	———	a_{2u}	a_{1g}	e_{2g}	———	e_{1g}	———
MY$_{10}$ pentagonal antiprism D_{5d}	a_{1g}	e_{1u}	———	a_{2u}	a_{1g}	e_{2g}	———	e_{1g}	———
MY$_{12}$ icosahedron I_h	a_g	t_{1u}	———	———	h_g	———	———	———	———

Table 8.5 Crystal Field Splitting Energies For Various Geometries[a]

	ρ = 1					ρ = 2				
Geometry	z^2	x^2-y^2	xy	xz	yz	z^2	x^2-y^2	xy	xz	yz
MY linear, $C_{\infty v}$	3.43	−1.43	−1.43	−0.285	−0.285	5.14	−3.14	−3.14	0.57	0.57
MY$_2$ linear, $D_{\infty h}$	6.86	−2.86	−2.86	−0.57	−0.57	10.28	−6.28	−6.28	1.14	1.14
Bent (90° angle),[b] C_{2v}	−0.43	4.43	−0.57	−1.715	−1.715	−2.14	6.14	1.14	−2.57	−2.57
MY$_3$ fac trivacant,[c] C_{3v}	3.00	3.00	−2.00	−2.00	−2.00	3.00	3.00	−2.00	−2.00	−2.00
Trigonal plane, D_{3h}	−0.65	2.895	2.895	−2.57	−2.57	−3.21	5.46	5.46	−3.85	−3.85
MY$_4$ tetrahedron, T_d	−2.67	−2.67	1.78	1.78	1.78	−2.67	−2.67	1.78	1.78	1.78
Square plane, D_{4h}	−0.86	8.86	−1.14	−3.43	−3.43	−4.28	12.28	2.28	−5.14	−5.14
Trigonal pyramid,[c] C_{3v}	2.78	1.465	1.465	−2.86	−2.86	1.93	2.32	2.32	−3.29	−3.29
Cis divacant, C_{2v}	6.43	1.57	−3.43	−2.285	−2.285	8.14	−0.14	−5.14	−1.43	−1.43
MY$_5$ pentagonal plane, D_{5h}	−1.07	4.825	4.825	−4.29	−4.29	−5.35	9.10	9.10	−6.42	−6.42
Trigonal bipyramid, D_{3h}	6.21	0.035	0.035	−3.14	−3.14	7.07	−0.82	−0.82	−2.72	−2.72
Square pyramid,[c] C_{4v}	2.57	7.43	−2.57	−3.715	−3.715	0.86	9.14	−0.86	−4.57	−4.57
MY$_6$ octahedron, O_h	6.00	6.00	−4.00	−4.00	−4.00	6.00	6.00	−4.00	−4.00	−4.00

Hexagonal plane, D_{6h}	−1.30	5.79	5.79	−5.14	−5.14	−6.42	10.92	10.92	−7.70	−7.70
Pentagonal pyramid,[e] C_{5v}	2.36	3.395	3.395	−4.575	−4.575	−0.21	5.96	5.96	−5.85	−5.85
MY_7 heptagonal plane, D_{7h}	−1.50	6.75	6.75	−6.00	−6.00	−7.49	12.74	12.74	−8.99	−8.99
Pentagonal bipyramid, D_{5h}	5.79	1.965	1.965	−4.86	−4.86	4.93	2.82	2.82	−5.28	−5.28
Hexagonal pyramid,[e] C_{6v}	2.13	4.36	4.36	−5.425	−5.425	−1.28	7.78	7.78	−7.13	−7.13
MY_8 cube, O_h	−5.34	−5.34	3.56	3.56	3.56	−5.34	−5.34	3.56	3.56	3.56
Square antiprism, D_{4d}	−5.34	−0.89	−0.89	3.56	3.56	−5.34	−0.89	−0.89	3.56	3.56
Octagonal plane, D_{8h}	−1.72	7.72	7.72	−6.86	−6.86	−8.56	14.56	14.56	−10.28	−10.28
Hexagonal bipyramid, D_{6h}	5.56	2.93	2.93	−5.71	−5.71	3.86	4.64	4.64	−6.58	−6.58
MY_{10} pentagonal antiprism, D_{5d}	−6.86	2.86	2.86	0.57	−10.28	6.28	6.28	−6.28	−1.14	−1.14
MY_{12} icosahedron, I_h	0.00	0.00	0.00	0.00	0.00	0.00	0.00	0.00	0.00	0.00

[a] z along figure axis except where noted.
[b] Ligands along x, y axes.
[c] Ligands along x, y, z axes.
[d] Ligands along x, y, ±z axes.
[e] $Y_{ax}MY_{ba}$ angle = 90°.

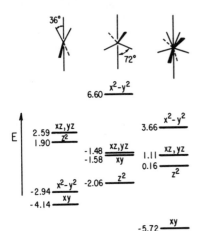

Figure 8.9 Derivation of dodecahedral MY_8 energy levels for $\rho = 1$. (Typical angles for MY_8 complexes are used.)

splitting energies for a selection of geometries of interest. The energies of the pentagonal D_{5h} MY_5 arrangement may be calculated in at least three different ways: (1) trigonal plane plus primary group 2, (2) 4 × primary group 2 minus trigonal plane, and (3) multiplication of the values for the trigonal plane by $\frac{5}{3}$. In the first two methods we need to average the xy, x^2-y^2 energies to produce the correct result. The third primary geometry (Table 8.3), D_{2d} MY_4, is useful in other circumstances and is the only one with a built-in angular dependence. We could for example determine the energies of the dodecahedral MY_8 system by the sum of two of these units (Figure 8.9).

Since in the icosahedral MY_{12} geometry all the d orbitals are equivalent, the CF splittings are zero. The pentagonal antiprism geometry (e.g., ferrocene), which is derived from the icosahedron by removal of the two axial ligands (Figure 8.10), has a splitting pattern (Table 8.5) that is just the inverse of that for linear MY_2. (The molecular orbital description of this molecule interposes other levels between these d orbitals, but the splitting pattern is similar.)

In general it is a relatively simple matter to generate the CF d-orbital energies for most high-symmmetry structures as in Table 8.5. If we wish to be strictly correct we would retain both Dq and Cp parameters and so be able to represent d-orbital energies more exactly on the model, but the present scheme has obvious attractions. The results may be used to plot

Figure 8.10 Derivation of pentagonal antiprism MY_{10} energy levels (e.g., ferrocene).

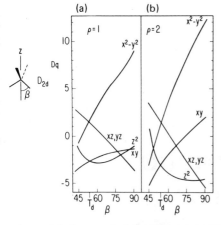

Figure 8.11 Correlation diagram for distortion of a tetrahedral. MY_4 molecule via a D_{2d} coordinate to a square planar structure; (a) and (b) were derived using values of the crystal field ρ parameter equal to 1 and 2, respectively.

out orbital correlation diagrams. From the results of Tables 8.3 and 8.5 we may trace the D_{2d} distortion of the tetrahedron to the square plane for both $\rho = 1$ and $\rho = 2$ (Figure 8.11). These give significantly different results. What is more difficult to do on Krishnamurthy and Schaap's approach is to examine for example the d-orbital energy changes involving geometry changes which cannot be synthesized using the third primary geometry, for example, the axial–basal angle variation in a C_{4v} MY_5 unit. The method is however readily applicable to the derivation of diagrams for many complexes containing different ligands, MY_5Y' for example.

$\rho = 2$	z^2	x^2-y^2
MY_6 octahedron	$6.00Dq(Y)$	$6.00Dq(Y)$
Less one Y group along z	$0.86Dq(Y)$	$9.14Dq(Y)$
Plus one Y' group along z	$0.86Dq(Y) + 5.14Dq(Y')$	$9.14Dq(Y) - 3.14Dq(Y')$

	xy	xz	yz
	$-4.00Dq(Y)$	$-4.00Dq(Y)$	$-4.00Dq(Y)$
	$-0.86Dq(Y)$	$-4.57Dq(Y)$	$-4.57Dq(Y)$
	$-0.86Dq(Y) - 3.14Dq(Y')$	$-4.57Dq(Y) + 0.57Dq(Y')$	$-4.57Dq(Y) + 0.57Dq(Y)$

There is a limitation on the sorts of mixed complex we may choose. When combining primary fragments all the ligands of the fragment must be the same, and the same requirement applies as before—no two d orbitals must transform as the same (nondegenerate) symmetry species. (Or no two pairs of d orbitals may transform as the same degenerate symmetry species.)

9 Theories of Transition Metal– Ligand Interaction: The Angular Overlap Approach

The AOM was specifically developed by Schäffer and Jørgensen[111,126, 176,185–189] for the case of transition metal–ligand interaction. We have however used it extensively to describe main group stereochemistry and properties earlier in this book. In its basic form only the quadratic term (e_λ or $\beta_\lambda S_\lambda^2$) is used and is sufficient to derive d-orbital energies which may be used analogously to the crystal field energies to describe the electronic spectroscopy and magnetic properties of transition metal species. In our discussion however we shall make use of the quartic term as well (f_λ or $\gamma_\lambda S_\lambda^4$) to distinguish several structural features.[23,25,27]

9.1 Derivation of Energy Diagrams

We are interested initially in a d orbital-only model just like the crystal field approach and we need to calculate the d-orbital interaction energies for a given geometry. The interaction of a central-atom p orbital with a ligand σ orbital gave a very simple angularly dependent function (Equation 1.22) but the d-orbital functions are a little more complex (Table 1.1). Four values which will be of great use are those for octahedrally based geometries and are shown in Figure 9.1. There is widespread evidence that σ interactions are usually stronger than π interactions, and so initially

142

Table 9.1 Transformation Properties of Ligand σ Orbitals in MY_n Species

MY	linear, C_{2v}	σ^+
MY_2	linear, $D_{\infty h}$	$\sigma_g^+ + \sigma_u^+$
	bent, C_{2v}	$a_1 + b_2$
MY_3	fac trivacant, C_{3v}	$a_1 + e$
	trigonal plane, D_{3h}	$a_1' + e'$
	T shape, C_{2v}	$2a_1 + b_2$
MY_4	tetrahedron, T_d	$a_1 + t_2$
	square plane, D_{4h}	$a_{1g} + b_{1g} + e_u$
	trigonal pyramid, C_{3v}	$2a_1 + e$
	cis divacant, C_{2v}	$2a_1 + b_1 + b_2$
MY_5	pentagonal plane, D_{5h}	$a_1' + e_1' + e_2'$
	trigonal bipyramid, D_{3h}	$2a_1' + e'$
	square pyramid, C_{4v}	$2a_1 + b_1 + e$
MY_6	octahedron, O_h	$a_{1g} + e_g + t_{1u}$
	hexagonal plane, D_{6h}	$a_{1g} + b_{1u} + e_{2g} + e_{1u}$
	pentagonal pyramid, C_{5v}	$2a_1 + e_1 + e_2$
MY_7	heptagonal plane, D_{7h}	$a_1' + e_1' + e_2' + 2_3'$
	pentagonal bipyramid, D_{5h}	$2a_1' + e_1' + e_2'$
	hexagonal pyramid, C_{6v}	$2a_1 + b_1 + e_1 + e_2$
MY_8	cube, O_h	$a_{1g} + t_{2g} + a_{2u} + t_{1u}$
	square antiprism, D_{4d}	$a_1 + b_2 + e_1 + e_2 + e_3$
	octagonal plane, D_{8h}	$a_{1g} + b_{1g} + e_{1u} + e_{2g} + e_{3u}$
	hexagonal bipyramid, D_{6h}	$2a_{1g} + a_{2u} + b_{1u} + e_{2g} + e_{1u}$
MY_{10}	pentagonal antiprism, D_{5d}	$a_{1g} + a_{2u} + e_{1g} + e_{1u} + e_{2g} + e_{2u}$
MY_{12}	icosahedron, I_h	$a_g + h_g + t_{1u} + t_{2u}$

143

Figure 9.1 Some useful values of metal d–ligand σ, π overlap integrals for octahedrally based geometries.

we concentrate on the σ manifold of orbitals. The high-symmetry octahedral case provides a good place to start. The ligand σ orbitals transform as $a_{1g} + e_g + t_{1u}$ (Table 9.1) and the metal d orbitals as $e_g + t_{2g}$ (Table 8.4). Thus we are interested only in MY interactions of species e_g. The symmetry-adapted ligand σ orbital combinations are readily found to be

$$\psi_{e_g}(1) = \frac{1}{2}(\phi_1 - \phi_2 + \phi_3 - \phi_4)$$

$$\psi_{e_g}(2) = \frac{1}{\sqrt{12}}(2\phi_5 + 2\phi_6 - \phi_1 - \phi_2 - \phi_3 - \phi_4)$$

(9.1)

$\psi(1)$ and $\psi(2)$ are of the correct form to overlap with x^2-y^2 and z^2, respectively. Use of Figure 9.1 gives

$$S_{e_g}(1) = \frac{1}{2} \cdot 4 \cdot \left(\frac{\sqrt{3}}{2}\right) S_\sigma$$

$$S_{e_g}(2) = \frac{1}{\sqrt{12}} \cdot 6 \cdot S_\sigma$$

(9.2)

both of which reduce to $S = \sqrt{3}S_\sigma$. Thus the interaction energy of the $e_g\sigma$ pair with z^2 and x^2-y^2 is $3\beta_\sigma S_\sigma^2 - 9\gamma_\sigma S_\sigma^4$ or $3e_\sigma - 9f_\sigma$ (from the basic AOM Equation 2.29). The result is shown pictorially in Figure 9.2, where we show two different e_σ and f_σ values for the bonding and antibonding orbitals. Clearly $\Delta_{oct} = 3e'_\sigma - 9f'_\sigma$. We may of course get an analogous

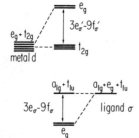

Figure 9.2 AOM approach to metal d–ligand σ interaction (octahedral molecule as example).

$$\psi\,(t_{2g}) = \tfrac{1}{2}\,(\phi_1 - \phi_2 + \phi_3 - \phi_4)$$ **Figure 9.3**   Ligand π-type orbital of t_{2g} symmetry.

result by using the ligand addivity Equation 2.35. For z^2 for example,

$$\epsilon = \beta_\sigma\left(1 + 1 + \frac{1}{4} + \frac{1}{4} + \frac{1}{4} + \frac{1}{4}\right)S_\sigma^2 = 3\beta_\sigma S_\sigma^2 \qquad (9.3)$$

The molecular orbital rationalization for the removal of the d-orbital degeneracy is now clear; the e_g set are involved in MY–σ antibonding interactions, the t_{2g} set are σ nonbonding. The size of Δ_{oct} thus increases with ligand σ strength.

Π bonding is readily included. The ligand π orbitals transform as t_{1g} + t_{2g} + t_{1u} + t_{2u} and we are therefore interested in interactions of t_{2g} symmetry. Figure 9.3 shows one component of the t_{2g} set. Clearly with reference to Figure 9.1,

$$S_{t_{2g}}(1) = \frac{1}{2}\cdot 4\cdot S_\pi \qquad (9.4)$$

and the interaction energy is $4e_\pi - 16f_\pi$ for each component. Whether the t_{2g} set is raised or lowered in energy depends upon whether the ligand π orbitals are acting as acceptors (e.g., CO, PR$_3$) or donors (e.g., halide) as in Figure 9.4. The MY σ bonding orbitals (mainly ligand located) have been left off for clarity. Hence the presence of π acceptors increases Δ_{oct} and that of π donors decreases Δ_{oct}. In the former case the t_{2g} set of d orbitals are MY π bonding; in the latter they are MY π antibonding.

For the tetrahedral coordination we may construct t_2 functions of σ symmetry with which to overlap xz, yz, and xy orbitals[79,117](Figure 9.5).

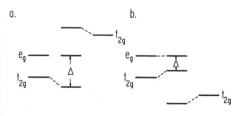

Figure 9.4   Effect of (a) π acceptors and (b) π donors on octahedral d-orbital energy levels in MY_6.

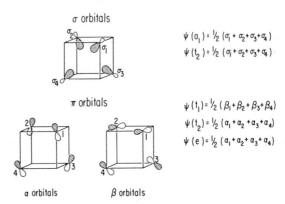

σ orbitals

$\psi\,(a_1) = \tfrac{1}{2}\,(\sigma_1 + \sigma_2 + \sigma_3 + \sigma_4)$

$\psi\,(t_2) = \tfrac{1}{2}\,(\sigma_1 + \sigma_2 + \sigma_3 + \sigma_4)$

π orbitals

$\psi\,(t_1) = \tfrac{1}{2}\,(\beta_1 + \beta_2 + \beta_3 + \beta_4)$

$\psi\,(t_2) = \tfrac{1}{2}\,(a_1 + a_2 + a_3 + a_4)$

$\psi\,(e\,) = \tfrac{1}{2}\,(a_1 + a_2 + a_3 + a_4)$

α orbitals   β orbitals

Figure 9.5 Ligand orbitals for a tetrahedral MY$_4$ geometry (only one component of each symmetry species is shown).

To evaluate the interaction energy we need to use the analytic form of the overlap integrals of Table 1.1. A shortcut method however exists since we know that all the σ interaction is contained in this representation and we may use the ligand sum rule (Equation 2.34) to give the total quadratic σ interaction energy as $4e_\sigma$. The quadratic interaction energy of each of the t_2 components is then $\tfrac{4}{3}e_\sigma$, and the total interaction $\epsilon = \tfrac{4}{3}e_\sigma - (\tfrac{4}{3})^2 f_\sigma$ $= \tfrac{4}{3}e_\sigma - \tfrac{16}{9}f_\sigma$. For the π-type orbitals there is no shortcut method available since both e and t_2 orbitals are involved in π-type interactions. Here we need to evaluate the π overlap integrals with the metal d orbitals (Figure 9.5). However once this has been done for one symmetry species (e.g., t_2) the ligand orbital sum rule may be used to calculate the interaction energy of the other symmetry set. The result is shown in Figure 9.6, where we show the case for π donors and have also assumed that the effect on the t_2 orbitals of σ and π interaction is an additive one. By comparison with the results for the octahedron,

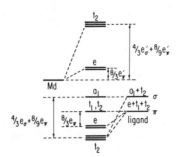

Figure 9.6 Angular overlap result for the tetrahedron.

$$\Delta_{tet} = \frac{4}{3}e_\sigma - \frac{16}{9}f_\sigma \pm \left(\frac{16}{9}e_\pi - \frac{512}{81}f_\pi\right)$$

$$\Delta_{oct} = 3e_\sigma - 9f_\sigma \pm (4e_\pi - 16f_\pi) \qquad (9.5)$$

$$\frac{4}{9}\Delta_{oct} = \frac{4}{3}e_\sigma - \frac{36}{9}f_\sigma \pm \left(\frac{16}{9}e_\pi - \frac{576}{81}f_\pi\right)$$

which shows that $\Delta_{tet} \simeq \Delta_{oct}$, the equality holds only if the quartic f terms are neglected. Using the overlap integrals of Figure 9.1 and this tetrahedral result, it is very easy to derive the d-orbital splittings of Figure 9.7. Table 9.2 shows σ and π interactions for a variety of geometries.

One advantage of the AOM over the CFT is that it is easy to derive the orbital energies in a coordination geometry containing an angular degree of freedom. For example in the C_{4v} square pyramid (Figure 9.8), from Table 1.1 we may readily show that

$$\epsilon(z^2) = \left[1 + 4\left(1 - \frac{3}{2}\sin^2\theta\right)^2\right]e_\sigma$$

$$\epsilon(x^2-y^2) = 4\left(\frac{\sqrt{3}}{2}\sin^2\theta\right)^2 e_\sigma$$

$$\epsilon(xz, yz) = 2\left(\frac{\sqrt{3}}{2}\sin 2\theta\right)^2 e_\sigma \qquad (9.6)$$

$$\epsilon(xy) = 0$$

plus the corresponding terms in f_σ. (Recall that the coefficient of the f term is just minus the square of the e term.) These we simply get from the ligand additivity equation and the square of the angular dependence of the overlap integral. The form of the Walsh diagram on bending the molecule then readily follows as in Figure 9.8. The less energetic π interactions may be simply added in if required.

We may similarly readily derive the orbital diagram for systems where two sets of d orbitals transform as the same symmetry species. Sometimes the answer is a trivial one. For example in the C_{3v} four-coordinate geometry (Figure 9.9) the ligand σ orbitals transform as $2a_1 + e$ and the d orbitals as $a_1 + 2e$. This is a straightforward three-orbital problem (there are a total of three a_1 orbitals and three pairs of e orbitals), and the middle orbital(s) of each set remain nonbonding. All the latent bonding interaction for a particular symmetry species is contained in the lowest-energy orbital and all the latent antibonding interaction is contained in the

Table 9.2 Angular Overlap d Orbital σ and π Interaction Energies for Various Geometries[a,b]

Geometry	z^2 e_σ	z^2 e_π	x^2-y^2 e_σ	x^2-y^2 e_π	xy e_σ	xy e_π	xz e_σ	xz e_π	yz e_σ	yz e_π
MY linear, $C_{\infty v}$	1	0	0	0	0	0	0	1	0	1
MY$_2$ linear, $D_{\infty h}$	2	0	0	0	0	0	0	2	0	2
bent (90° angle), C_{2v}	$\frac{1}{2}$	0	$\frac{3}{2}$	0	0	2	0	1	0	1
MY$_3$ fac trivacant, C_{3v}	$\frac{3}{2}$	0	$\frac{3}{2}$	0	0	2	0	2	0	2
trigonal plane, D_{3h}	$\frac{3}{4}$	0	$\frac{9}{8}$	$\frac{3}{2}$	$\frac{9}{8}$	$\frac{3}{2}$	0	$\frac{3}{2}$	0	$\frac{3}{2}$
T-shape, $C_{2v}{}^{d}$	$\frac{3}{2}$ —$\frac{3}{2}$	0	$\frac{3}{2}$ —$\frac{3}{2}$	0	0	2	0	3	0	1
MY$_4$ tetrahedron, T_d	0	$\frac{8}{3}$	0	$\frac{8}{3}$	$\frac{4}{3}$	$\frac{8}{9}$	$\frac{4}{3}$	$\frac{8}{9}$	$\frac{4}{3}$	$\frac{8}{9}$
square plane, D_{4h}	1	0	3	0	0	4	0	2	0	2
trigonal pyramid, $C_{3v}{}^{e}$	$\frac{7}{4}$	0	$\frac{9}{8}$	0	$\frac{9}{8}$	$\frac{3}{2}$	0	$\frac{5}{2}$	0	$\frac{5}{2}$
cis divacant, C_{2v}	$\frac{5}{2}$	0	$\frac{3}{2}$	0	$\frac{9}{8}$	2	0	3	0	3
MY$_5$ pentagonal plane, D_{5h}	$\frac{5}{4}$	0	$\frac{15}{8}$	$\frac{3}{2}$	$\frac{15}{8}$	$\frac{5}{2}$	0	$\frac{5}{2}$	0	$\frac{5}{2}$
trigonal bipyramid, D_{3h}	$\frac{11}{4}$	0	$\frac{9}{8}$	$\frac{3}{2}$	$\frac{9}{8}$	$\frac{3}{2}$	0	$\frac{7}{2}$	0	$\frac{7}{2}$
square pyramid, $C_{4v}{}^{e}$	2	0	3	0	0	4	0	3	0	3

MY$_6$ octahedron, O_h	3	0	3	0	0	4	0	4
hexagonal plane, D_{6h}	$\frac{3}{2}$	0	3	$\frac{9}{4}$	$\frac{9}{4}$	3	0	3
pentagonal pyramid, C_{5v} [e]	$\frac{9}{4}$	0	$\frac{9}{4}$	$\frac{11}{8}$	$\frac{11}{8}$	$\frac{5}{2}$	0	$\frac{7}{2}$
MY$_7$ heptagonal plane, D_{7h}	$\frac{7}{4}$	0	$\frac{21}{4}$	$\frac{21}{8}$	$\frac{21}{8}$	$\frac{7}{2}$	0	$\frac{7}{2}$
pentagonal bipyramid, D_{5h}	$\frac{13}{4}$	0	$\frac{15}{8}$	$\frac{15}{8}$	$\frac{15}{8}$	$\frac{5}{2}$	0	$\frac{9}{2}$
hexagonal pyramid, C_{6v} [e]	$\frac{5}{2}$	0	$\frac{7}{4}$	$\frac{7}{4}$	$\frac{7}{4}$	3	0	4
MY$_8$ cube, O_h	0	$\frac{4}{3}$	0	$\frac{8}{3}$	$\frac{8}{3}$	$\frac{16}{9}$	$\frac{8}{3}$	$\frac{16}{9}$
square antiprism, D_{4d} [g]	0	$\frac{8}{3}$	$\frac{4}{3}$	$\frac{4}{3}$	$\frac{4}{3}$	$\frac{32}{9}$	$\frac{8}{3}$	$\frac{37}{9}$
octagonal plane, D_{8h}	2	0	3	3	3	4	0	4
hexagonal bipyramid, D_{6h}	$\frac{2}{5}$	0	$\frac{9}{4}$	$\frac{9}{4}$	$\frac{9}{4}$	3	0	5
MY$_{10}$ pentagonal antiprism, D_{5d} [h]	$\frac{12}{5}$	$\frac{24}{5}$	$\frac{12}{5}$	$\frac{12}{5}$	$\frac{12}{5}$	$\frac{24}{5}$	$\frac{12}{5}$	$\frac{14}{5}$
MY$_{12}$ icosahedron, I_h	$\frac{12}{5}$	$\frac{24}{5}$	$\frac{12}{5}$	$\frac{12}{5}$	$\frac{12}{5}$	$\frac{24}{5}$	$\frac{12}{5}$	$\frac{24}{5}$

[a] The coefficients p_σ and p_π are given for the quadratic interaction energy $\epsilon = p_\sigma e_\sigma + p_\pi e_\pi$.
[b] z along figure axis except where noted.
[c] Ligands along x, y, z.
[d] z^2, x^2-y^2 mix together in this point group.
[e] $Y_{ax}MY_{ba} = 90°$.
[f] Ligands along x, y, $\pm z$.
[g] MY bond makes angle of 57.4° with z-axis.
[h] Ligands in same position as for icosohedron.

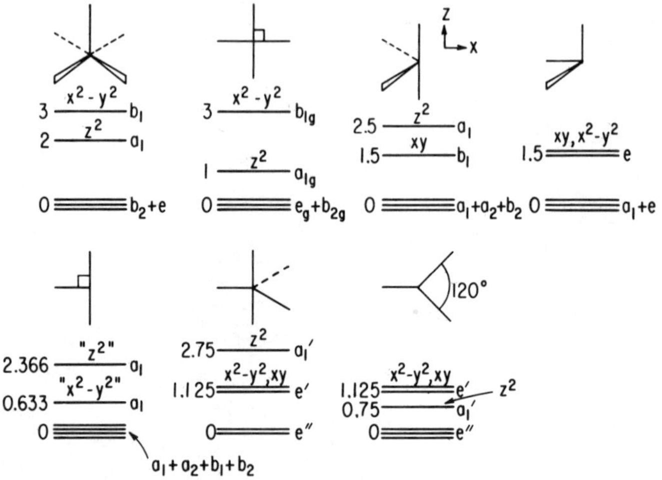

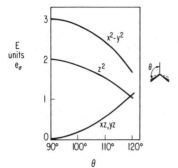

Figure 9.7 AOM d-orbital splittings for some simple MY_n geometries of interest using a σ-only model. Only the quadratic part of the interaction energies is shown (units of e_σ).

highest-energy orbital. The actual description of the d orbitals is however not so simple. For the e species interaction for example the d orbitals are described as a mixture of xz, yz and x^2-y^2, xy. The actual mixing coefficient varies with geometry. Here the middle nonbonding orbital is a metal d orbital. In the case of the a_1 orbitals the nonbonding orbital of the three-orbital problem is ligand-located. In the absence of ligand–ligand interactions it will remain unchanged in energy when metal d–ligand σ interaction is switched on.

For the T-shaped molecule however the σ orbitals transform as $2a_1 + b_1 + b_2$ and the d orbitals as $2a_1 + a_2 + b_1 + b_2$. Here we need to solve a secular determinant[135] for the a_1 orbitals. The diagonal terms are

Figure 9.8 Walsh diagram for distortion within the C_{4v} dinate of an MY_5 system using the AOM.

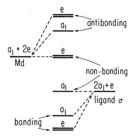

Figure 9.9 AOM molecular orbital diagram for a C_{3v} MY$_4$ species showing the nonbonding metal e levels and nonbonding ligand σ a_1 orbital.

given in the normal way by $\Sigma\, e_\lambda F_\lambda^2(n)$ as in Equation 2.35, but the off-diagonal term which controls the mixing between ϕ_1 and ϕ_2 is given by

$$\sum_n e_\lambda F_\lambda(n, \phi_1) F_\lambda(n, \phi_2). \qquad (9.7)$$

Here ϕ_1 and ϕ_2 are the two d orbitals of the same symmetry species, in this case z^2 and x^2-y^2 (Figure 9.10). In the present case[135] the diagonal elements are $(1 + 2 \times \frac{1}{4})e_\sigma$ and $(2 \cdot \frac{3}{4})e_\sigma$ and the off diagonal elements are $[\frac{\sqrt{3}}{2} \cdot (-\frac{1}{2}) + \frac{\sqrt{3}}{2} \cdot (-\frac{1}{2}) + 0 \cdot 1]e_\sigma$, which gives Equation 9.8 for the d-orbital interaction energies:

$$\begin{vmatrix} 1.5e_\sigma - E & -\dfrac{\sqrt{3}}{2}e_\sigma \\[3mm] \dfrac{\sqrt{3}}{2}e_\sigma & 1.5e_\sigma - E \end{vmatrix} = 0 \qquad (9.8)$$

This has roots

$$E = \left(1.5 \pm \frac{\sqrt{3}}{2}\right)e_\sigma \qquad (9.9)$$

The result is shown schematically in Figure 9.10. For many of the systems

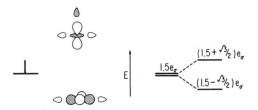

Figure 9.10 Interaction energies of z^2 and x^2-y^2 in the C_{2v} T-shape molecule where they both transform as a_1. Without allowing mixing between them, the interaction energies are both $1.5e_\sigma$. On mixing they split apart.

considered in this chapter, two d orbitals transform as the same symmetry species; but because of the geometry choice (usually octahedrally based), ligand overlap with one of these is zero. Thus the octahedral cis-divacant geometry may be quite simply treated. Although z^2 and x^2-y^2 both transform as a_1, the overlap of x^2-y^2 with the ligand σ orbitals in the xy plane is zero for the axis choice of Figure 9.7.

There are some interesting similarities between the CFT using Krishnamurthy and Schaap's method and the AOM (Table 9.3). Such similarities between the AOM and CFT are to be expected. The AOM replaces the destabilization of a d orbital by a ligand charge with a σ antibonding interaction with that d orbital. Taking each point in turn:

1. There is a sum rule for the quadratic d-orbital interaction energies (in terms of e_σ). These are energetically superior to the f_σ terms for which a similar rule does not apply. The total crystal field splitting energy of all five d orbitals is zero.
2. The d-orbital energies are geometrically additive on both schemes as long as mixing between two d orbitals does not occur.
3. This is clearly related to (2).
4. In principle we could include e_δ the use of which is reasonably rare.

There are also the f_λ terms, but since much of the structural and spectroscopic data is understandable without them, we have not included them in this comparison. (Recall that their coefficients are also related since the interaction energy associated with a particular symmetry may be written as $\epsilon = pe_\lambda - p^2 f_\lambda$). In cubic geometries only one CFT parameter

Table 9.3 Similarities Between Crystal Field and Angular Overlap Approaches to Transition Metal–Ligand Interaction

Crystal Field	Angular Overlap
a. Σ Crystal field splitting energy $= 0$	$\Sigma F_\lambda^2 = n_\lambda$
b. Crystal field splitting energy of a given orbital in ML_n = sum of splittings in $ML_{n1} + ML_{n2}$ ($n = n_1 + n_2$)	Interaction energy of a given orbital with the ligands in ML_n = sum of interaction energies in $ML_{n1} + ML_{n2}$ ($n = n_1 + n_2$)
c. Total splitting energy of in-plane orbitals independent of orientation of in plane ligands	Total interaction energy of in plane orbitals independent of orientation of in plane ligands
d. Two parameters Dq, Cp.	Two parameters e_σ, e_π

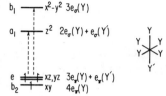

Figure 9.11 Quantitative estimation of ligand σ donor strength e_σ by the spectral analysis of C_{4v} MY_5Y' systems. Only the transitions shown are electric dipole allowed. The difference in orbital energy between b_1 and a_1 orbitals is the $\Delta\sigma$ of Table 9.4.

(Dq) is needed, but two AOM parameters are needed to describe the d orbital splittings. Of the two methods the AOM is the more versatile since we can derive diagrams (a) as a function of an angular degree of freedom and (b) in low-symmetry situations which the simple approach of Krishnamurthy and Schaap is unable to do (for these cases we have to revert to a full CFT treatment).

Table 9.4 Values of e_σ and e_π From a Series of Metal Complexes (Units 1000 cm^{-1})

		$e_\sigma{}^a$	$e_\pi{}^a$	$\Delta\sigma^b$
Cr(III)	CN$^-$			1.31
	OH$^-$	8.67	3.0	
	NH$_3$	7.03 $\pm$ 0.02	0	0
	N̲CS$^-$			-1.0
	H$_2$O	7.9 $\pm$ 0.6		-1.1
	N$_3^-$			-1.36
	F$^-$	7.39 $\pm$ 0.07	1.69 $\pm$ 0.03	-1.41
	Cl$^-$	5.54 $\pm$ 0.14	1.16 $\pm$ 0.14	-2.12
	Br$^-$	4.92 $\pm$ 0.09	0.83 $\pm$ 0.39	-2.51
	I$^-$			-2.97
	py	5.85 $\pm$ 0.09	$-0.67 \pm$ 0.07	
Pa(IV)	F$^-$	2.866	1.230	
	Cl$^-$	1.264	0.654	
	Br$^-$	0.976	0.683	
	I$^-$	0.725	0.618	
U(V)	F$^-$	4.337	1.792	
	Cl$^-$	2.273	1.174	
	Br$^-$	1.775	1.174	

[a] Ligand–d orbital interactions in Cr(III), Ref. 76. Ligand–f orbital interactions in Pa(IV) and U(V), Ref. 214.

[b] Ligand–d orbital interactions $\Delta\sigma$ values (the energy difference between b_1 and a_1 orbitals of Figure 9.10),Ref. 166, for Cr(III)(NH$_3$)$_5$Y species.

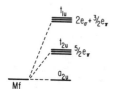

Figure 9.12   Splitting of metal f orbitals in an octahedral MY_6 system. The e_σ and e_π parameters apply to metal f orbital–ligand σ, π interactions.

The method is equally applicable to the study of complexes with mixed ligands. Figure 9.11 shows a diagram derived for an MY_5Y' species using the ligand additivity method. By the analysis of a large number of spectra of this sort of molecule and $C_{4v}MY_4Y'Y''$ species, numerical values may be obtained for the e_σ and e_π parameters. (As we noted above, the e_λ parameters only have been used in spectral analysis. The quartic f_λ terms are neglected.) Table 9.4 shows the best values and their estimated standard deviations for a series of ligands coordinated to Cr(III).[76] Because the electronic spectral results give values for the energy differences between octahedral t_{2g}-derived and e_g-derived levels, the data do not provide an absolute energy scale. Values are therefore shown relative to $e_\pi(NH_3)$ = 0. Figures are also given for the e_σ and e_π parameters obtained by an analysis of f-orbital complexes[214] where, since the f orbitals split into three components under O_h (Figure 9.12), the two observed f^1 transitions give values for the AOM parameters without recourse to the analysis of the spectra of lower-symmetry molecules.

9.2   The Meaning of AOM Parameters[70]

Recall that for the metal d orbitals the parameter governing their energy is

$$e_\lambda = \frac{H_{YY}^2 S_\lambda^2}{H_{MM} - H_{YY}} \tag{9.10}$$

where H_{YY} and H_{MM} are the ligand λ and the metal d-orbital ionization potentials. Just as we examined the variation of the CFT parameters Cp and Dq as a function of M and Y (Section 8.3), we may do the same for the AOM by making reference to a table of experimental data collected by Gerloch and Slade[70] (Table 9.5).

1. As the overlap integral decreases (usually as the bond length increases), the e_λ will decrease. π bonding is more sensitive to changes in bond length, and so e_π will change faster.

Table 9.5 Angular Overlap Parameters From Electronic Spectral Data
(Units 1000 cm^{-1})

	Equatorial Ligands Y			Axial Ligands Y'		
	e_σ	e_π	e_σ/e_π	e_σ	e_π	e_σ/e_π
D_{4h} MY$_4$Y$_2'$ Complex						
Ni pyridine$_4$Cl$_2$	4.670	0.570	8.19	2.980	0.540	5.52
Ni pyrazole$_4$Cl$_2$	5.480	1.370	4.00	2.540	0.380	6.68
Ni pyridine$_4$Br$_2$	4.500	0.500	9.00	2.540	0.340	7.21
Ni pyrazole$_4$Br$_2$	5.440	1.350	4.03	1.980	0.240	8.25
Cr(en)$_2$F$_2^+$	7.233	—	—	8.033	2.000	4.02
Cr(en)$_2$(H$_2$O)$_2^{3+}$	7.833	—	—	7.497	1.410	5.32
Cr(en)$_2$Cl$_2^+$	7.500	—	—	5.857	1.040	5.63
Cr(en)$_2$Br$_2^+$	7.500	—	—	5.120	0.750	6.83
D_{2d} MY$_4$ Complex						
CuCl$_4^{2-}$	6.764	1.831	3.69			
CuBr$_4^{2-}$	4.616	0.821	5.62			

Adapted from Ref. 70.

2. An increase in $|H_{MM}|$ as the effective metal nuclear charge increases leads to an increase in e_λ.

3. An increase in $|H_{YY}|$ as the electronegativity of the ligand increases leads to an increase in e_λ.

For the cupric halides CuX$_4^{-2}$ in Table 9.5, on going from X = Cl to X = Br we may examine the effects of these three factors: (1) The overlap integrals decrease so both e_λ decrease ($\lambda = \sigma, \pi$) but e_σ/e_π increases. (2) Z_{eff} decreases, so again e_λ decreases and e_σ/e_π increases. (3) Chlorine is more electronegative than bromine, so the e_λ decrease again.

Similar arguments apply to the Cren$_2$X$_2$ complexes, and the experimental data are understandable if the electronegativity of O lies between that of F and Cl. For the Nipy$_4$Cl$_2$ and Nipyrazole$_4$Cl$_2$ complexes a longer Ni–Cl bond is found for the pyrazole complex. This shows up as a smaller e_σ and e_π but larger e_σ/e_π ratio for M–Cl interaction in the latter system in agreement with experiment.

Perhaps the largest difference between this discussion and the earlier one for the CFT (Section 8.3) is that the e_σ and e_π values are in line with simple ideas of chemical bonding. The AOM has a more physically realistic basis than the CFT although the qualitative success achieved above should not mask the fact that the whole basis of the approach is based on a very crude molecular orbital model.

9.3   Some Restrictions on the Model

The AOM gives the d-orbital energies on a d-orbital-only model. In some situations there may be other central atom orbitals mixed into the molecular orbitals we usually call "d orbitals." (A brief analysis of this topic forms the basis for Chapter 13.) In the tetrahedral MY_4 geometry the metal d orbitals transform as $e + t_2$ and the p orbitals as t_2. Thus d–p mixing allowed by symmetry is to be expected in such molecules and is indeed responsible for the much larger (by a factor of 10^2) intensities of "d–d transitions" in tetrahedral geometries compared to their octahedral analogs.[61] When the tetrahedral complex is distorted to square planar, then d–s mixing only is allowed and d–p mixing is forbidden by symmetry (all p orbitals are u; d and s orbitals are g). This d–s mixing occurs between z^2 and s, and the result is depression of the z^2 orbital in energy (Figure 9.13). This shows up in the electronic spectra of these species as follows.[92] There are three possible intrasystem electronic transitions for low spin (ls) d^8 and d^9 species (in orbital terms transitions from (xz, yz), xy and z^2 to x^2–y^2) and only two angular overlap parameters, e_σ and e_π. For $d^9\text{Cu(DACO)}_2(\text{ClO}_4) \cdot 2H_2O$ (DACO = diazacyclooctane) and the ls $d^8\text{Ni(II)}$ analog, the use of the two transitions originating in π-type orbitals leads to $e_\sigma = 7800 \text{ cm}^{-1}$, $e_\pi = 1300 \text{ cm}^{-1}$ ($e_\sigma/e_\pi = 6.0$) and $e_\sigma = 10870$ cm^{-1}, $e_\pi = 1885 \text{ cm}^{-1}$ ($e_\sigma/e_\pi = 5.8$), respectively. However, whereas the transition from z^2 to x^2–y^2 should occur at about two thirds the energy of that from (xz, yz) to x^2–y^2, it is found clustered together with these transitions and so occurs at a higher energy than expected (Figure 9.13). This depression of z^2 is quite striking, about 5000 to 6000 cm^{-1} for these Ni(II) and Cu(II) DACO complexes and represents an approximate mixing coefficient of 0.2 for the $4s$ orbital into the orbital we call z^2. EPR spectra of square planar ls d^7 Co(II) complexes with one electron in z^2 indicate, via the size of the isotropic hyperfine coupling constant, that the electron spends about 3 to 6% of the time in the $4s$ orbital. This also indicates a mixing coefficient of about 0.2. The d orbital-only model does have to be viewed and used with care in these systems, but the modifications we need to introduce (e.g., s- and p-orbital mixing) are usually readily understandable in molecular orbital terms.

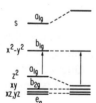

Figure 9.13   Effect of metal d–s orbital mixing on the electronic transition arising via promotion of an electron from z^2 to x^2–y^2.

As we mentioned above, one way the AOM parameters have been evaluated quantitatively for Cr(III) is by an analysis of the spectra of Cr(III)$Y_4Y'Y''$ species (Y, Y', Y'' = F^-, Cl^-, Br^-, NH_3, py, OH^-, H_2O), and the results are given in Table 9.4. A related analysis measured the difference in the orbital contributions to the two marked transitions of Figure 9.11 and thus also provided a way to rank the ligands Y in a σ strength order in the complexes $Cr^{III}(NH_3)_5Y$ (Y = NH_3 or H_2O). The values of $\Delta\sigma$ are also shown in Table 9.3. There are quantitative differences between the two sets of data, but the order of σ strength for the halogens is clearly $F^- > Cl^- > Br^-$ ($>I^-$). A similar result applies to the data from f-orbital complexes. This halogen order is the opposite to that usually envisaged. I^- is usually considered as being the best σ donor and F^- the worst. However we must remember that these AOM parameters are obtained from electronic spectral studies and are concerned with the MY antibonding orbitals. Other assessments of σ donor strength are concerned with the properties of the MY bonding orbitals which are occupied. In fact the e_σ values for the two orbitals are different since from Equation 2.28.

For antibonding orbitals

$$e'_\sigma = \frac{H^2_{YY}S^2_{MY}}{H_{YY} - H_{MM}}$$

For bonding orbitals $\qquad\qquad\qquad\qquad\qquad\qquad\qquad\qquad$ (9.11)

$$e_\sigma = \frac{H^2_{MM}S^2_{MY}}{H_{YY} - H_{MM}}$$

Thus the order of the AOM e_σ parameters associated with a series of ligands for the bonding orbitals is governed by changes in $S^2_{MY}/(H_{YY} - H_{MM})$ since H_{MM} is constant for all the systems with a given metal atom. But the ordering of the e'_σ values for the antibonding orbitals is weighted in addition by the H^2_{YY} term. This is $\frac{16}{9}$ times larger for F^- than for Cl^-, for example (using the VSIP values for the ns orbitals from Table 2.1) and may well be responsible for the reversal of the e'_σ order compared to the expected e_σ order. Caution is then needed when using σ strength parameters obtained via electronic spectra to view ground-state properties controlled by occupation of bonding orbitals.

10 Crystal Field and Molecular Orbital Stabilization Energy

A very useful set of results which may be readily extracted from the CFT and AOM values of the d-orbital energies are the stabilization energies associated with a given arrangement of n d electrons in the five d orbitals relative to the free atom or ion plus ligands.[24,53,69,105,156] This is particularly interesting to compute as a function of the d-electron configuration, and very often molecular properties may be rationalized purely on the basis of electronic configuration rather than the exact nature of the metal and the ligands.

10.1 CFSE and MOSE

We need to recognize that, since the five d orbitals are usually non-equivalent in coordination complexes, there are often several ways to arrange the electrons. As a simple example Figure 10.1 shows three configurations of five d electrons which we call low, intermediate, and high spin (ls, is, and hs). We define the orbital occupation numbers of these three configurations as 22100, 21110, and 11111, respectively. These give the number of electrons in each d orbital starting with the lowest-energy one first. The orbital energy increases in the order ls < is < hs, but the two-electron terms in the energy decrease in this order so that the two effects compete against each other for the lowest energy arrangement. The two-electron terms are larger in the low-spin case since here there is a loss in exchange energy when the spins are paired and also an increase in Coulomb repulsion energy when the two electrons reside in the same orbital (Table 10.1). The Racah interelectronic repulsion parameters are

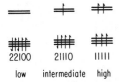

Figure 10.1   Orbital occupation diagrams and numbers for low-, intermediate-, and high-spin d^5 octahedral MY_6 molecules.

a useful combination of relevant two-electron integrals of the type shown in Equation 2.6, which describe the coulombic and exchange interactions between electrons located in metal d orbitals. In terms of these parameters the low-spin condition for d^4-d^7 systems is met for the following series of inequalities:

$$d^4: \quad \Delta > 6B + 5C$$

$$d^5: 2\Delta > 15B + 10C$$

$$d^6: 2\Delta > 5B + 8C$$

$$d^7: \quad \Delta > 4B + 4C$$

The "pairing energy" (P) is defined such that high-spin complexes are found for $\Delta < P$ and low-spin complexes for $\Delta > P$. Figure 10.2 shows this diagrammatically. The intermediate spin system is never stable with respect to either the high- or low-spin systems, and indeed it is very rarely found. Table 10.1 shows some mean pairing energies for a series of first-row metal ions, and, combined with Table 8.1, a prediction may be made as to the likely spin state for a given system. At the boundary of stability two spin states may be in equilibrium, and a change of temperature or pressure might be sufficient for their interconversion.[134]

Table 10.1   Pairing Energies for First-Row Ions^a

Configuration	Pairing Energy Contributions			Experimental (1000 cm^{-1})	
	Coulomb	Exchange	Total	M^{2+}	M^{3+}
d^4	$2C$	$6B + 3C$	$6B + 5C$	23.5	28.0
d^5	$B + 2C$	$6.5B + 3C$	$7.5B + 5C$	25.5	30.0
d^6	$-B + 2C$	$3.5B + 2C$	$2.5B + 4C$	17.6	21.0
d^7	$2C$	$4B + 2C$	$4B + 4C$	19.5	23.5

^{a} Well-authenticated intermediate spin systems are represented only by (distorted) octahedral chlorophthalocyanato Fe(III) and halobis(diethyldithiocarbamato) iron(III).

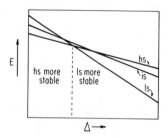

Figure 10.2 Variation in stability of low-, intermediate-, and high-spin octahedral MY_6 molecules as a function of Δ, the e_g–t_{2g} orbital energy separation. The crossover occurs at $\Delta = P$.

The stabilization energy of a complex assembled from ligands and metal atom or ion depends dramatically upon whether it is viewed from the CFT or AOM standpoints. Figure 10.3 shows schematically the difference between the two schemes. The crystal field stabilization energy (CFSE) is the energy change for a given arrangement of d electrons on going from a spherically symmetrical environment of charges to a nonspherical one. For the octahedral geometry that is used as an illustration, each electron in the t_{2g} orbital is stabilized by $4Dq(2\Delta/5)$, and each electron in the e_g orbital is destabilized by $6Dq(3\Delta/5)$. Thus the CFSE is given in general by equation 10.1 as a function of n, the number of electrons:

$$\text{CFSE} = n(t_{2g}) \cdot \frac{2}{5}\Delta - n(e_g) \cdot \frac{3}{5}\Delta \qquad (10.1)$$

Thus the CFSEs for the three configurations of Figure 10.1 are 2Δ, Δ, and 0, respectively. On this scheme the d^0 configuration has a zero CFSE. By using the molecular orbital approach however the molecular orbital stabilization energy (MOSE) is the energy change associated with occupied d orbitals on forming the complex (Figure 10.3) plus the stabilization of the four ligand σ electrons and, for this d^0 configuration, is certainly nonzero. If we assume for the sake of simplicity that the e_σ values representing the bonding and antibonding orbitals are equal, then, on a σ only model, ignoring the less important f_σ term,

$$\text{MOSE} = 12e_\sigma - 3n(e_g)\,e_\sigma \qquad (10.2)$$

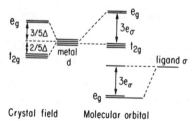

Figure 10.3 Comparison of CFT and AOM approaches to the "stabilization energy" (octahedral system used as an illustration) of a particular d-orbital configuration.

For the three configurations of Figure 10.1, the MOSEs are $12e_\sigma$, $9e_\sigma$, and $6e_\sigma$, respectively. Equation 10.3 gives a general shortcut to the evaluation of the MOSE or $\Sigma(\sigma)$, the total σ stabilization energy in any geometry:

$$MOSE = \sum_i h_i \epsilon_i \qquad (10.3)$$

Since occupied d orbitals produce a destabilizing contribution which exactly (on our model) cancels the stabilization energy of the corresponding bonding orbital, the only bonding orbitals which contribute to total stabilization energy $\Sigma(\sigma)$ are those whose antibonding counterpart is wholly or partially unoccupied. Here h_i describes the number of electron holes (0, 1, 2) in the ith d orbital and ϵ_i is its MY interaction energy.

10.2 Thermodynamic Properties

For the enthalpy changes associated with the reaction of Equation 10.4 for first-row transition metal ions,

$$M_{gas}^{2+} + 6H_2O_{gas} \rightarrow M(H_2O)_6^{2+} \qquad (10.4)$$

a gradual increase on moving from left to right is expected. A similar gradation in other properties such as the lattice energies of MX_2 species or MX bond length (related to the "ionic radius" of M) is perhaps also to be anticipated as the number of d electrons increases. In fact none of these plots is smooth (Figure 10.4). In all of these examples the actual form of the diagram is a double-humped curve. This phenomenon is ascribed to the effect of the ligand field which splits the d orbitals apart and leads to some electronic configurations receiving a larger stabilizing effect than others. Thus the smooth curves expected for these plots apply only to the energy changes associated with a spherical ligand potential, and for metal complexes we need to add the CFSE and MOSE (depending upon whether we are using the CFT or AOM) associated with each configuration. This is not a monotonic function of d^n.

Figure 10.5 shows the d-orbital contributions to the energy of the system on the two schemes. The forms of the two plots derived from Equations 10.1 and 10.2 are quite different but with the addition of a sloping background (Figure 10.6) they produce identical results. The CFT explanation of the extra contribution to ΔH_{hyd}^0, or the lattice energy is that the electrostatic interaction between the M^{x+} ion and the n negatively charged

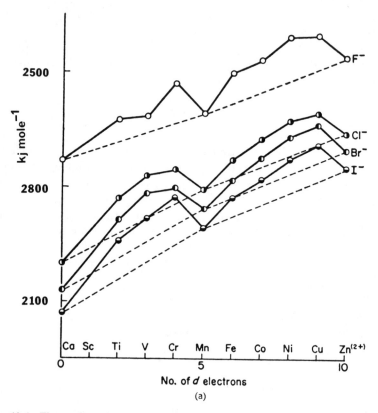

Figure 10.4 Thermodynamic and structural properties of octahedral $M^{II}Y_6$ species as a function of d-electron configuration. Species are in the high-spin configuration: (*a*) lattice energies of first-row MY_2 species; (*b*) lattice energy of first-row oxides, sulfides, and selenides; (*c*) heats of hydration of first-row divalent (open circles) and trivalent ions (full circles); (*d*) metal–ligand distances in first-row divalent metal-containing systems; (*e*) standard metal–divalent ion electrode potentials (open circles: data corrected for ionization potential variations; solid circles: uncorrected data). Reprinted with permission from *Introduction to Ligand Fields*, by B. N. Figgis, Wiley, New York (1966).

ligands of the MY_n complex increases across the series as the MY distance decreases.

The AOM explanation is quite different but goes beyond the d orbital-only model. The a_{1g} and t_{1u} ligand orbitals (Figure 9.2) are of the correct symmetry to interact with $(n + 1)s$ and p orbitals located on the central atom, respectively. The result will be a stabilization contribution extra to that supplied by metal d-ligand interaction. A complete σ molecular orbital diagram is shown in Figure 10.7. As the first transition metal series is traversed, the metal ionization potential increases and the metal–ligand

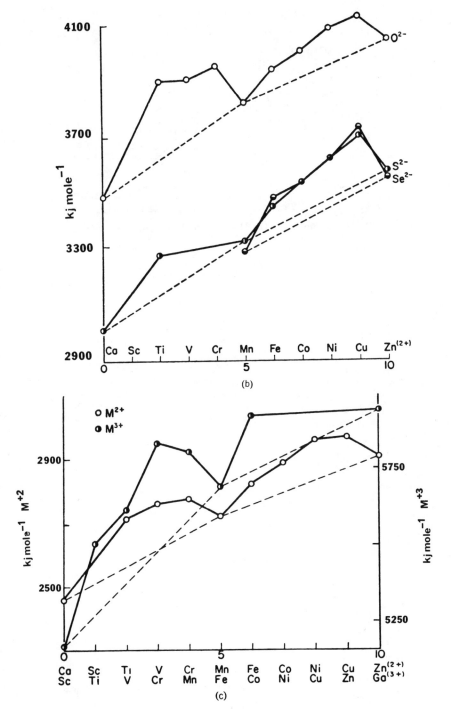

Figure 10.4 (Continued)

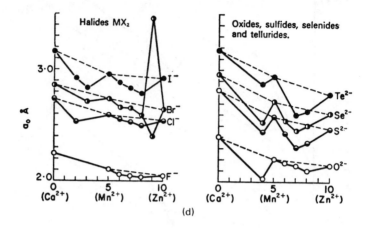

(d)

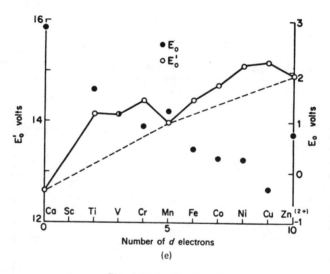

(e)

Figure 10.4 (Continued)

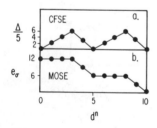

Figure 10.5 Variation in (a) crystal field stabilization energy (CFSE) and (b) molecular orbital stabilization energy (MOSE) as a function of d^n for high-spin octahedral $M^{II}Y_6$ species.

164

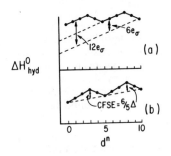

Figure 10.6 Plots of (a) and (b) of Figure 10.5, but with a sloping background added to mimic the observed behavior in Figure 10.4.

distance shortens. With increasing ionization potential the energy separation between $(n + 1)s$ and p orbitals and the ligand σ orbitals decreases. The overall effect is a larger stabilization energy from this source at the right-hand side of the first-row series than at the left. (Going even further to the right, Group III B onward, we normally only consider s and p orbitals in bonding.) The plot of Figure 10.6 compared with those of Figure 10.4 allows an assessment of the importance of the d-orbital contribution compared to that via interaction with s and p orbitals. For the lattice energies of the MX_2 molecules (X = I) for example, the d-orbital contribution is about 20% for Ti^{2+}, 10% for Co^{2+} and Mn^{2+}, and of course 0% for Zn^{2+}. (Strictly speaking we ought to allow for different values of e_σ (different Δ) for the different transition metal ions in any quantitative comparisons.) For X = F the d-orbital contribution is somewhat smaller. Inclusion of π bonding alters these results slightly. For the case of π donors (Figure 9.4), recalling that all the MY π bonding orbitals are filled,

$$MOSE = 12e_\sigma + 24e_\pi - 4n(t_{2g})\,e_\pi - 3n(e_g)\,e_\sigma \qquad (10.4)$$

e_π is usually less than e_σ, and in many cases $e_\pi/e_\sigma \sim 0.25$. Making this assumption,

$$MOSE = 18e_\sigma - e_\sigma[n(t_{2g}) + 3n(e_g)] \qquad (10.5)$$

This does not change the overall appearance of the curve but leads to a small increase in the calculated importance of the metal d orbitals in our

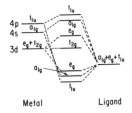

Figure 10.7 A σ-only molecular orbital diagram for an octahedral MY_6 complex showing ligand interactions with central-atom s and p orbitals. (No significance should be attached to the relative ordering of the bonding molecular orbitals.)

a.

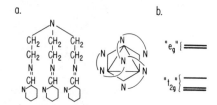

b.

Figure 10.8 (*a*) The ligand trenpy and its co-ordination of a metal atom. (*b*) Schematic molecular orbital diagram for approximate octahedral coordination by six of the N atoms of the metal atom and the capping of this structure by the seventh N atom of the ligand.

examples above. In fact, although we have assumed that e'_σ and e_σ of Figure 10.3 are equal for pedagogic reasons, the same qualitative behavior is predicted if in fact they are different. In this case the origin of the sloping background contains an extra factor, the variation in e_σ across the series. An interesting difference between the crystal field and molecular orbital approaches is the smaller importance of the *d*-orbital contribution in the ionic model (Figure 10.6) to the heat of formation and other thermodynamic properties.

The variation in M–O or M–S bond lengths in the oxides, sulfides, and so on (Figure 10.4*c*) as a function of d^n is rationalized along similar

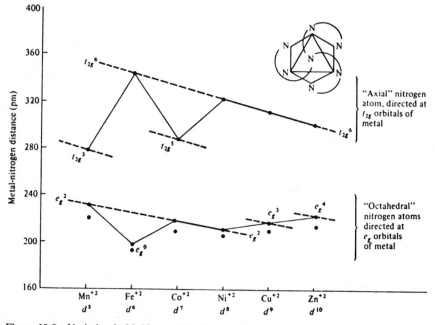

Figure 10.9 Variation in M–N$_{oct}$ and M–N$_{cap}$ as a function of the *d*-electron configuration for M(trenpy). Reprinted with permission from J. E. Huheey, *Inorganic Chemistry*, 2nd ed., Harper and Row, New York (1978).

lines. The larger the MOSE or CFSE, the shorter the M–O distance. One very interesting example that illustrates this effect involves the ligand N(CH$_2$CH$_2$N=CHC$_5$H$_4$N)$_3$ which forms [102] a very similar series of complexes with all the first-row metals Mn^{2+} to Zn^{2+}. Six of the N atoms coordinate to the metal in an approximately octahedral geometry, and one is coordinated via a face of the octahedron (Figure 10.8). The six octahedral ligands are able to interact in σ fashion with the z^2 and x^2–y^2 orbitals in an exactly analogous way to that described above. The capping ligand however interacts with one of the t_{2g} orbitals on the metal giving rise to the approximate molecular orbital diagram of Figure 10.8b. Thus occupation of "e_g" orbitals leads to destabilization and hence lengthening of the octahedral MN linkages, whereas occupation of the highest-energy "t_{2g}" orbital leads to lengthening of the MN capping or "axial" ligand bond. In Figure 10.9 there is a large difference in MN$_{cap}$ distance between $(t_{2g})^6$ and $(t_{2g})^5$ points in the Fe^{2+} and Co^{2+} cases as expected on the scheme and a large change in the MN$_{oct}$ distance for Fe^{2+} compared to the systems with $(e_g)^2$.

Another process that can be viewed[53,169] using the change in CFSE or MOSE as a function of d^n is the variation in rate constant for loss of H$_2$O from the hexaquo ions:

$$M(H_2O)_6^{2+} \overset{k_1}{\rightleftharpoons} M(H_2O)_5^{2+} + H_2O \qquad (10.6)$$

It is assumed that the rate-controlling feature is the activation enthalpy $\Delta H^{\ddagger}$ for ligand loss:

$$K_1 = A \exp\left(\frac{-\Delta H^{\ddagger}}{RT} + \frac{\Delta S^{\ddagger}}{R}\right) \propto \exp\frac{-\Delta H^{\ddagger}}{RT} \qquad (10.7)$$

This may be calculated as a function of d^n by computing the difference in CFSE or MOSE for the process of Figure 10.10. (The CFT and AOM splitting patterns for octahedral and spy geometries are shown in Figures 8.2 and 9.7, respectively.) Since the latter is noncubic, Dq and Cp parameters are needed to describe the energy levels. From Table 8.5 we have extracted the relevant figures for two different ratios of $a^2\langle r^2\rangle/\langle r^4\rangle$ = ρ. Table 10.2 shows the differences in CFSE for ρ = 1, 2 and MOSE for the ligand loss. In Figure 10.11 these differences are plotted against d^n and compared with the observed log k_1 plot. From Equation 10.7 and

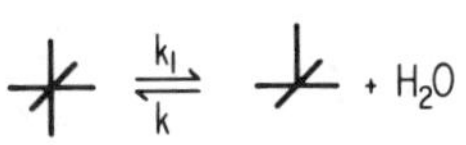

Figure 10.10 Loss of H$_2$O from octahedral M^{II}(H$_2$O)$_6^{2+}$.

Table 10.2 Difference in MOSE and CFSE Between the Octahedral MY_6 and Square Pyramidal MY_5 Units for High-Spin Systems

Electronic Configuration	MOSE			CFSE[a] (Units of Dq)		
	Octahedron	Square Pyramid	Difference	Octahedron	Square Pyramid	Difference
d^0	$12e_\sigma + 24e_\pi$	$10e_\sigma + 20e_\pi$	$2e_\sigma + 4e_\pi$	0	0 (0)	0
d^1	$12e_\sigma + 20e_\pi$	$10e_\sigma + 17e_\pi$	$2e_\sigma + 3e_\pi$	-4.0	-3.72 (-4.57)	0.28 (-0.57)
d^2	$12e_\sigma + 16e_\pi$	$10e_\sigma + 14e_\pi$	$2e_\sigma + 2e_\pi$	-8.0	-7.43 (-9.14)	0.57 (-1.14)
d^3	$12e_\sigma + 12e_\pi$	$10e_\sigma + 10e_\pi$	$2e_\sigma + 2e_\pi$	-12.0	-10.0 (-10.0)	2.0 (2.0)
d^4	$9e_\sigma + 12e_\pi$	$8e_\sigma + 10e_\pi$	$e_\sigma + 2e_\pi$	-6.0	-7.43 (-9.14)	-1.43 (-3.14)
d^5	$6e_\sigma + 12e_\pi$	$5e_\sigma + 10e_\pi$	$e_\sigma + 2e_\pi$	0	0 (0)	0 (0)
d^6	$6e_\sigma + 8e_\pi$	$5e_\sigma + 7e_\pi$	$e_\sigma + e_\pi$	-4.0	-3.72 (-4.57)	0.28 (-0.57)
d^7	$6e_\sigma + 4e_\pi$	$5e_\sigma + 4e_\pi$	e_σ	-8.0	-7.43 (-9.14)	0.57 (-1.14)
d^8	$6e_\sigma$	$5e_\sigma$	e_σ	-12.0	-10.0 (-10.0)	2.0 (2.0)
d^9	$3e_\sigma$	$3e_\sigma$	0	-6.0	-7.43 (-9.14)	-1.43 (-3.14)
d^{10}	0	0	0	0	0 (0)	(0)

[a] Values for $\rho = 2$ in parentheses.

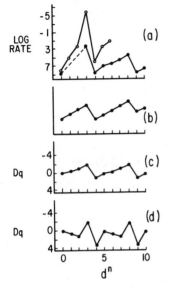

Figure 10.11 Observed values of log k_1 plotted (a) as a function of d^n for the reaction of Figure 10.10; full circles for M^{2+} ions, open circles for M^{3+} ions. Calculated ΔMOSE results (b) and calculated ΔCFSE results for $\rho = 1$ (c) and $\rho = 2$ (d) for this reaction. A sloping background has been added to the results in (b) to mimic the observed rate dependence.

assuming $\Delta H^{\ddagger} = \Delta$CFSE (or ΔMOSE) $+ \Delta'$, where Δ' is the non-d-orbital contribution,

$$\log k_1 \propto \Delta\text{CSFE (or MOSE)} + \Delta' \qquad (10.8)$$

A sloping background has been added as in Figure 10.6, but here it represents the variation of Δ'. The correlation between theory and experiment is quite good using the AOM; d^3 and d^8 complexes are "inert" (low rate constants) while the others are described as "labile" (higher rate constants). For the CFSE plots, the one for $\rho = 2$ is not a very good match with experiment.[125] For $\rho = 1$ the fit is similar to the MOSE approch.

However spectroscopic data on square pyramidal $M(H_2O)_5^{2+}$ species of this sort are not available to experimentally determine a value of ρ. For the MOSE plot of Figure 10.11, similar results are obtained if $e_\pi = 0$ (σ-only model) or $e_\pi/e_\sigma \simeq 0.25$.

11 Geometries of Transition Metal Complexes

In general for main group structures the observed geometry for a given species was close to that expected by comparison with isoelectronic molecules. The exceptions were of interest in themselves—the planar nature of $N(SiH_3)_3$ and $C(CN_3)^-$ or the octahedral geometry of $TeBr_6^{2-}$. By way of contrast transition metal complexes at first sight present a much less coherent picture. First the structures seem much more flexible. For example, applying pressure to crystals of $CuCl_4^{2-}$ or $Ni(CN)_5^{3-}$ results in reversible geometry changes.[60] Changing the solvent of crystallization or the temperature or other experimental variables results in different structurs for the Lifshitz salts—yellow and diamagnetic (square planar with two extra weakly coordinated ligands) or blue and paramagnetic (octahedral). Changing the matrix medium allows observation of two different geometries (C_{3v} and D_{2d}) for the unstable d^9 species $Co(CO)_4$.

Second, "unusual" geometries and coordination numbers may be forced on molecules by either chelating or bulky ligands (cf. $N(SiH_3)_3$).[55] For example trigonal prismatic six coordination may be achieved by suitable ligand choice even though the octahedral geometry is sterically and usually electronically favored for comparable systems with monodentate ligands. (In solid-state systems containing transition metals the coordination environment around the metal very often reflects that found in molecular systems.[34,218] Square planar coordination, a feature of Pt(II) chemistry, is found in the structure of solid PtO.)

Last, for main group molecules the number of possible electronic configurations was comparatively small (two to seven pairs of valence electrons) with half-pairs (odd numbers of electrons) being relatively rare, and high- and low-spin species essentially limited to two-coordinate carbon. By way of contrast transition metal complexes range through 11 basic configurations (d^0-d^{10}) usually with more than one arrangement of the electron spins available for each (high, intermediate, or low).

11.1 VSEPR Ideas

We note that the VSEPR approach does not work for transition metal species. The problem is how to accommodate the d electrons on the scheme. In some cases (for example metal carbonyls) it has been suggested[71] that the d electrons are involved in π bonding and so may be neglected in the VSEPR count. Thus the tetrahedral, trigonal bipyramidal, and octahedral geometries of $Ni(CO)_4(d^{10})$, $Fe(CO)_5(d^8)$, and $Cr(CO)_6(d^6)$, respectively, are rationalized on the basis of four, five, and six σ pairs, respectively. However for unsaturated carbonyls non-VSEPR geometries are often found. Figure 11.1 shows some results from matrix isolation experiments[23] where these unstable molecules may be studied. The D_{3h} geometry which the VSEPR model does predict is found for unsaturated d^{10} systems ML_3, where M = Ni, Pd, or Pt; L = CO or N_2. Including all the d electrons does not improve matters. $Cr(CO)_3$ (total six pairs) could be described as being based on an octahedron (Figure 11.2), but (a) the CCrC angles are greater rather than less than 90°, and (b) the ligand arrangement maximizes lone pair–lone pair repulsions. In addition $Fe(CO)_3$, with a total of six pairs plus two half-pairs, has a similar structure. Similarly although $Rh(PPh_3)_3^+$ (seven pairs) is a T-shape molecule[224]

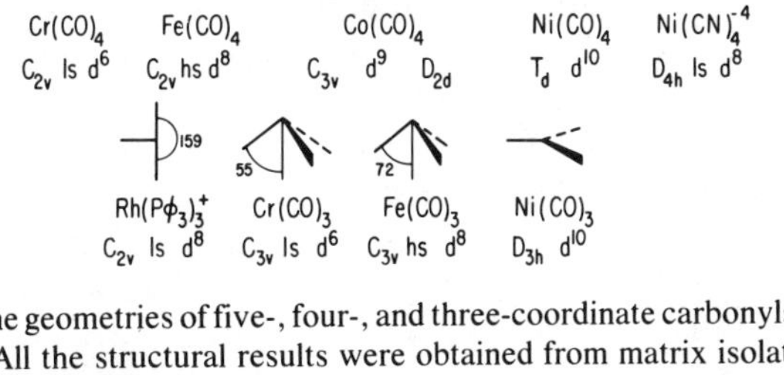

Figure 11.1 Some geometries of five-, four-, and three-coordinate carbonyls, and $Ni(CN)_4^{2-}$ and Rh $(P\phi_3)_3^+$. All the structural results were obtained from matrix isolation experiments (except for $Fe(CO)_5$, $Ni(CO)_4$, $Ni(CN)_4^{2-}$, and $Rh(P\phi_3)_3^+$). These structures are representative of the geometries adopted by these species as a function of d-electron configuration.

Figure 11.2 "Prediction" of the fac trivacant structure for $Cr(CO)_3$ and the T-shape structure of $Rh(PPh_3)_3{}^+$ using the VSEPR approach based on an octahedron and pentagonal bipyramid of electron pairs, respectively.

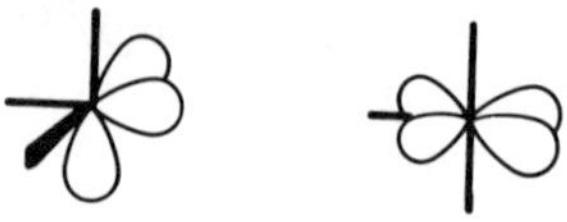

which is in accord with a pentagonal bipyramid VSEPR structure (again with maximum lone pair–lone pair repulsions), octahedral OsF_6 also has seven pairs. An alternative means to view these geometries is clearly needed.

11.2　The Jahn–Teller Approach

The (true or first-order) Jahn–Teller theorem states that a nonlinear orbitally degenerate molecule will distort so as to remove this degeneracy. The result is based on the group theoretical properties of the term containing S_i in Equation 5.3, which describes via perturbation theory the energy dependence of a system (for convenience the electronic ground state Ψ_0) as it is distorted along a coordinate S_i. From group theory the symmetry species of S_i is contained in the symmetrical direct product of the symmetry species of Ψ_0. Table 11.1 lists the allowed species of the distortion coordinate for orbitally degenerate parent molecules. Many aspects of structural transition metal chemistry are traditionally viewed in this way. Current usage of the theorem may be divided into two parts: (a) recognition that in most point groups, asymmetric occupation of degenerate orbitals leads to the possibility of a degenerate electronic state; (b) the observation that for cis MY_4Y_2' species, for example, distorted geometries are observed similar to those found in MY_6 species, even though in the substituted molecules the point symmetry (C_{2v} in this case) is not high enough to support a degenerate representation.

In both of these cases we often need to go a little further than the original result of Jahn and Teller to view structural distortions. In Figure 11.3 are shown three examples containing asymmetric occupation of degenerate orbitals. Only in case a of the figure is a degenerate electronic ground state unambiguously predicted which may be viewed using the first order Jahn–Teller theorem. In case d, to determine whether the ^1E_g or the $^1A_{1g}$ state lies lower in energy requires a knowledge of the two-electron terms in the energy. (Tanabe–Sugano diagrams of transition metal systems in cubic fields always find the degenerate state(s) to be lower in energy than the nondegenerate A_{1g} state arising via the direct product.) In case c of Figure 11.3, which applies to e^2 configurations in molecules of particular point groups containing fourfold axes of symmetry, no degenerate electronic states result. In both cases c and d we

Table 11.1 Jahn–Teller Allowed Vibrations and Point Groups Accessible via Operation of a Jahn–Teller Distortion[a]

Parent Point Group	Jahn–Teller Active Vibrations	Electronic State Split	Ground-State Symmetries Consistent with the Operation of a Jahn–Teller Distortion[c]
O_h	e_g	$E_g, E_u, T_{1g}, T_{1u}, T_{2g}, T_{2u}$	D_{4h}, D_{2h} (rhombus)
	t_{2g}	$T_{1g}, T_{1u}, T_{2g}, T_{2u}$	D_{3d}, D_{2h} (rectangle) C_{2h}, C_i
T_d	e	E, T_1, T_2	D_{2d}, D_2
	t_2	T_1, T_2	C_{3v}, C_{2v}, C_s, C_1
T_h	e_g	E_g, E_u, T_g, T_u	D_{2h}
	t_g	T_g, T_u	C_{2h}, S_6, C_i
O	e	E, T_1, T_2	D_4, D_2
	t_2	T_1, T_2	D_3, D_2, C_2, C_1
T	e	E, T	D_2
	t	T	C_3, C_2, C_1
D_{6h}	e_{2g}	$E_{1g}, E_{2g}, E_{1u}, E_{2u}$	D_{2h}, C_{2h}
D_{4h}	b_{1g}	E_g, E_u	D_{2h} (rhombus)
	b_{2g}	E_g, E_u	D_{2h} (rectangle)
D_{3h}	e'	E', E''	C_{2v}, C_s
C_{6h}	e_{2g}	$E_{1g}, E_{1u}, E_{2g}, E_{2u}$	C_{2h}
C_{4h}	$2b_g$	E_g, E_u	C_{2h}
C_{3h}	e'	E', E''	C_s
C_{6v}	e_2	E_1, E_2	C_{2v}, C_2
C_{4v}	b_1	E	C_{2v}
	b_2	E	C_{2v}
C_{3v}	e	E	C_s, C_1
D_{3d}	e_g	E_g, E_u	C_{2h}, C_i
D_{2d}	b_1	E	D_2
	b_2	E	C_{2v}
D_6	e_2	E_1, E_2	D_2, C_2
D_4	b_1	E	D_2
	b_2	E	D_2
D_3	e	E	C_2, C_1
C_6	e_2	E_1, E_2	C_2
C_4	$2b$	E	C_2
C_3	e	E	C_1

Table 11.1 (Continued)

S_6	e_g	E_g, E_u	C_1
S_4	$2b$	E	C_2
I_h	g_g	G_g, G_u, H_g, H_u	$T_h, D_{3d}, C_{2h}, S_6, C_i$
	$2h_g$	$T_{1g}, T_{1u}, T_{2g}, T_{2u}, G_g, G_u, H_g,$ H_u	$D_{5d}, D_{3d}, D_{2h}, C_{2h}, C_i$
I	g	G, H	T, D_3, C_3, C_2, C_1
	$2h$	T_1, T_2, G, H	D_5, D_3, D_2, C_2, C_1
$D_{\infty h}$	None[b]		
D_{5h}	e'	E_2', E_2''	C_{2v}, C_s
	e_2'	E_1', E_1''	C_{2v}, C_s
C_{5h}	e_1'	E_2', E_2''	C_s
	e_2'	E_1', E_1''	C_s
$C_{\infty v}$	None[b]		
C_{5v}	e_1	E_2	C_s, C_1
	e_2	E_1	C_s, C_1
D_{6d}	b_1	E_3	D_6
	b_2	E_3	C_{6v}
	e_2	E_1, E_5	D_2, C_{2v}, C_2
	e_4	E_2, E_4	D_{2d}, S_4
D_{5d}	e_{1g}	E_{2g}, E_{2u}	C_{2h}, C_i
	e_{2g}	E_{1g}, E_{1u}	C_{2h}, C_i
D_{4d}	b_1	E_2	D_4
	b_2	E_2	C_{6v}
	e	E_1, E_3	D_2, C_{2v}, C_2
D_5	e_1	E_2	D_2, C_1
	e_2	E_1	C_2, C_1
C_5	e_1	E_2	C_1
	e_2	E_1	C_1
S_{12}	$2b$	E_3	C_6
	e_2	E_1, E_5	C_2
	e_4	E_2, E_4	S_4
S_{10}	e_{1g}	E_{2g}, E_{2u}	C_i
	e_{2g}	E_{1g}, E_{1u}	C_i
S_8	$2b$	E_2	C_4
	e_2	E_1, E_3	C_2

[a] Adapted from Ref. 112.
[b] Linear molecules for which the effect does not hold (see Ref. 19).
[c] Only point groups accessible by small displacements are included. Thus a tetrahedral molecule on distortion may proceed via the D_{2d} route eventually to the square planar D_{4h} structure (not included in the table).

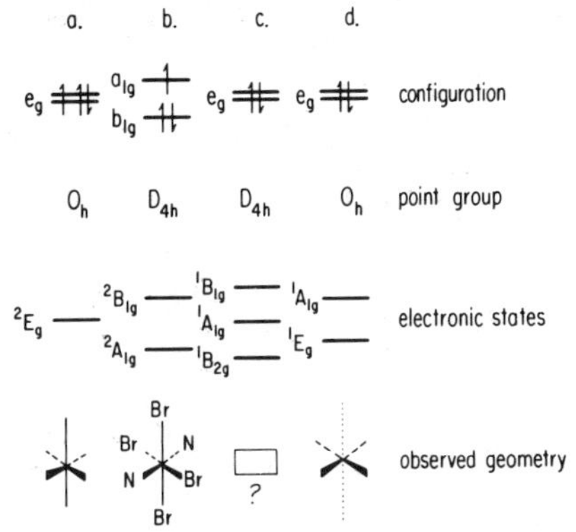

Figure 11.3 Some examples of molecules referred to as "Jahn–Teller unstable": (a) the $(e_g)^3$ configuration, the "classic" transition metal case of Cu(II) in an octahedral environment; (b) as (a) but where the presence of a different coordination sphere (in this case 4Br$^-$ and 2NH$_3$) around the copper atom reduces the maximum local symmetry from O_h to D_{4h}; (c) the cyclobutadiene molecule, where, even though there is asymmetric occupation of degenerate orbitals, the symmetry species of the electronic ground state is not degenerate; (d) low-spin octahedral M(II) (M = Ni, Pd, Pt) species which do not exist since two axial ligands are detached and square planar units result.

need to consider the S_i^2-dependent term in Equation 5.3 to understand the distortion. The second term in brackets allows a stabilization of the system if the states Ψ_0 and Ψ_m (one of the higher energy states generated by the particular electron configuration) mix together on distortion. Group theory tells us that this may occur if the direct product of the symmetry species of Ψ_0 and Ψ_m contains the symmetry species of S_i. Alternatively knowledge of the symmetry species of the electronic states derived from a given electronic configuration allows prediction of the symmetry species of S_i. This higher-order variant of the Jahn–Teller effect where the states which couple together on distortion are separated in energy by two-electron energy terms is described as the pseudo Jahn–Teller effect.

Case b of Figure 11.3 is an example of (b) above. Here in the D_{4h} point group the E_g state of the regular octahedron has split into two. Which lies to lower energy in the undistorted geometry will be determined by the ligand properties. These two states may couple together on distortion in a similar fashion to that described for cases c and d. The term second-order Jahn–Teller effect is used to describe this process where the two states involved are separated initially in energy by different one-electron (orbital) terms in the energy (in addition to perhaps differences in the two-electron terms). In Chapter 5 the use of this approach to view main group stereochemistry was particularly useful. In orbital terms the second-order Jahn–Teller effect is exactly equivalent to the second-order mixing of molecular orbitals on distortion. The two orbitals (usually HOMO and LUMO), of different symmetry in the parent undistorted molecule, become of the same symmetry species in the distorted geometry. In this way it differs from the first-order and pseudo Jahn–Teller results where the two components of the initially degenerate orbitals transform as different symmetry species in the distorted structure. In general the form of the coordinate S_i predicted via the second-order effect in these substituted molecules is very similar to that of one of the components of the degenerate distortion mode predicted via application of the first-order effect in the unsubstituted parent.

One must be careful however not to ascribe distorted structures as a result of operation of the Jahn–Teller theorem without some care. For example the molecules $MX_3(NMe_3)_2$ where M(III) = Ti, V, or Cr, have idealized D_{3h} geometry but are all distorted in a C_{2v} fashion. The size of the distortion increases in the order $V(d^2) < Ti(d^1) < Cr(d^3)$. The larger distortion for the Cr than for the Ti analog is understandable on the basis of a predicted static $(d^3, e''^2 e'^1 \equiv {}^4E')$ compared to a dynamic $(d^1, e''^1 \equiv {}^2E'')$ distortion, respectively, for these species as we shall see later. The distortion of the vanadium complex is not explicable on Jahn–Teller grounds at all $(d^2, e''^2 \equiv {}^3A_2')$ and the suspicion does arise that perhaps factors other than Jahn–Teller effects are responsible for the distortions of the Ti and Cr complexes too. Similarly for the series of related molecules of Figure 11.4, in case a the first-order Jahn–Teller result is inapplicable since the structure is a linear one but the second-order approach, using higher-energy orbitals as outlined in Chapter 5, may be used to predict a nonlinear structure. However cases b (isomorphous with the cyclobutadiene problem) and c may be in principle tackled using the Jahn–Teller recipe described above. (Again cis structures are predicted.) Case d, which chemically is little different from the others, may not employ Jahn–Teller ideas. The two orbitals derived from the degenerate π-type orbitals in N_2F_2 do not mix together on distortion as expected from the second-order approach and typified by case d as shown in Figure 11.3e and f. They are orthogonal in both linear and bent geometries. This suggests that the first-order and pseudo Jahn–Teller effects associated

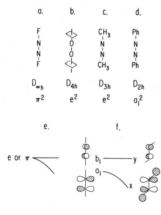

Figure 11.4 Some A_2X_2 molecules which are bent at both A atoms. (a) N_2F_2 which exists as cis and trans isomers; N_2H_2 probably has similar isomeric forms; (b) the $(NH_3)_5$ Co OO Co$(NH_3)_5{}^{+4}$ ion isoelectronic with N_2X_2 (W. P. Schaefer, *Inorg. Chem.* **7**, 725 (1968); (c) azomethane and (d) azobenzene which exist as cis and trans isomers; (e) behavior of the degenerate π-type orbitals of A_2X_2 on distortion; (f) exactly analogous behavior of these π-type orbitals, as in azobenzene when these are not degenerate.

with asymmetric degenerate orbital occupancy in cases b and c are not the real reason behind the observation of nonlinear skeletons in these molecules but perhaps a second-order Jahn–Teller or orbital mixing effect involving higher-energy orbitals as demonstrated for N_2F_2. With these qualifications concerning the Jahn–Teller theorems in mind we shall see how the geometries of simple coordination compounds may be viewed in general by the application of Jahn–Teller ideas to systems with asymmetric occupation of degenerate orbitals.

It is found experimentally that the "size" of the "Jahn–Teller distortion" varies widely in the transition metal systems. Sometimes it is very small and the molecule exists in a dynamic state around some equilibrium geometry. Cooling the sample leads to freezing out of the distortion. This "dynamic Jahn–Teller" effect most often occurs when the orbital degeneracy is associated with an unsymmetrical arrangement of electrons in the σ nonbonding orbitals (but involved perhaps in π interactions) of the complex. A static Jahn–Teller distortion usually occurs when the unsymmetrical electronic arrangement involves occupation of the MY σ antibonding orbitals.

The molecules MF_6 (M = Mo–Rh or W–Pt) are all regular octahedral molecules[217] in the gas phase, although some of them are predicted to be Jahn–Teller unstable (Table 11.2). The effect associated with an orbital degeneracy involving the t_{2g} orbitals shows up as vibronic coupling in their spectra. The octahedral geometry for six coordinated ligands is clearly the best steric arrangement for this system. Two alternative geometries, the bicapped tetrahedron and trigonal antiprism, are less satisfactory in this regard. $V(CO)_6$ is a low-spin d^5 octahedral carbonyl. Its gas-phase electronic spectrum is considerably broader than that for the ls d^6 (and Jahn–Teller stable molecule) $Cr(CO)_6$ and indicates a dynamic system. The crystal structure of the molecule at low temperatures shows[16] a small distortion (Figure 11.5). The crystal structure of the hs d^6 molecule $FeCl_4^{2-}$, $(t_2)^4(e)^2$, shows it to be a regular tetrahedron within experimental error. Exceptions to this generalization, where small or imperceptible distortions are found in molecules with orbital degeneracy of the σ antibonding levels, are found for a few systems. At room temperature

Table 11.2 Transition Metal Hexafluorides

Jahn–Teller Stable	Jahn–Teller Unstable
d° MoF_6 d° WF_6	d^1 ReF_6 d^2 RuF_6
d^3 RhF_6 d^3 IrF_6	d^2 OsF_6 d^4 PtF_6

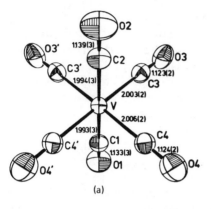

Figure 11.5 Molecular structure of (*a*) V(CO)$_6$ (reprinted with permission from S. Bellard, K. A. Rubinson, and G. M. Sheldrick, *Acta Crystallogr.* **B35** 271 (1979) and (*b*) K$_2$BaCu(NO$_2$)$_6$ (reprinted with permission from S. Takagi, P. G. Lenhert, and M. D. Joesten, *J. Am. Chem. Soc.* **96** 6608 (1974). Copyright by the American Chemical Society).

(a)

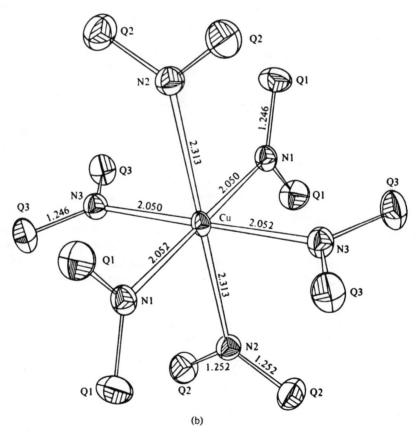

(b)

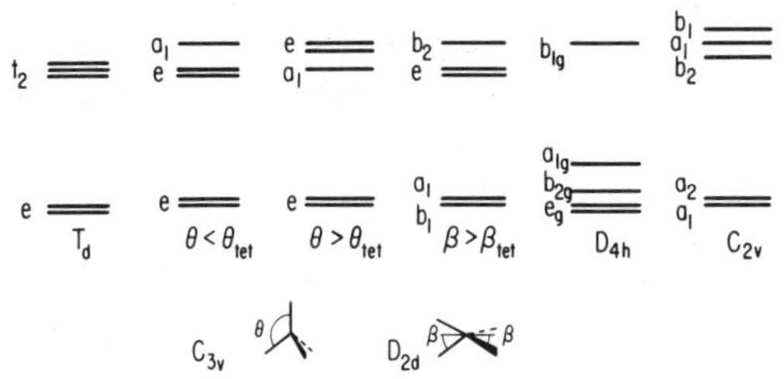

Figure 11.6 Qualitative energetic behavior of the metal d orbitals on distorting the tetrahedral MY_4 structure.

$K_2PbCu(NO_2)_6$, containing $Cu^{II}(NO_2)_6^{4-}$ ions, appears undistorted to the X-ray technique. The crystal is cubic and the copper atom is octahedrally coordinated by six N atoms. However, cooling the crystal by 10 degrees (to 276 K) reveals a static distortion.[199] Alternatively replacement of Pb by Ca or Ba leads to a distorted structure (Figure 11.5). The NiX_4^{2-} ion (hs d^8) is another species, predicted to be distorted but found in some crystalline environments as a regular tetrahedron.

For the static distortions seen in molecules it is helpful to have a series of qualitative d-orbital splitting patterns available (Figure 11.6) in order to follow the operation of the effect more fully. We use the criterion that the actual geometry must not contain unsymmetrical occupation of the highest orbitals (those involved in σ bonding) and begin with a survey of four-coordinate structures. The d^{10} molecule is Jahn–Teller stable at the tetrahedral reference geometry, which is favored on steric grounds and is of course the VSEPR structure for four σ pairs. Tetrahedral molecules with this electron configuration include $Ni(CO)_4$ and $ZnCl_4^{2-}$. The tetrahedral d^9 molecule is Jahn–Teller unstable. It has a $^2T_2$ electronic state which according to the symmetry arguments of the first-order Jahn–Teller approach may be split via an e or t_2 motion. Although group theory does not prescribe the phase of the distortion we may see from Figure 11.6 that the degeneracy is only removed for $\theta < \theta_{tet}$ in the C_{3v} distortion coordinate. $Co(CO)_4$ has this structure with $\theta \sim 100°$ when synthesized in a CO matrix at 20 K. CuX_4^{2-} (X = Cl or Br) however usually distort via the D_{2d} route, and the geometry may go all the way to planar if a small cation (e.g., NH_4^+) is chosen, or by application of high pressure.

d^8 complexes come in two types, high and low spin (22211 and 22220, respectively). We shall not dwell on the factors stabilizing one configuration with respect to another but ask how the two molecules might distort. For both low- and high-spin cases arguments analogous to those above apply. The molecule should distort from tetrahedral. Low-spin d^8 molecules are always found to be square planar, for example,

$Ni(CN)_4^{2-}$, and the rationalization of this geometry follows along lines similar to those for the d^9 system. High-spin d^8 molecules are clearly unstable at the tetrahedral and squashed D_{2d} geometries. C_{3v} ($\theta > \theta_{tet}$), D_{2d} ($\beta < \beta_{tet}$), and C_{2v} are open to relieve the Jahn–Teller instability. The C_{2v} structure is found for $Fe(CO)_4$, but isoelectronic NiX_4^{2-} species are often undistorted when viewed by X-ray crystallography. High-spin d^7 (22111) species are stable at the tetrahedral geometry, and $CoCl_4^{2-}$ has this structure. Low-spin d^7 systems are Jahn–Teller unstable at this geometry. From the set of orbital diagrams of Figure 11.6 the square planar or C_{2v} geometries are open to this configuration. Square planar structures are always found. The d^6 species may exist as low (22200), intermediate (22110), or high (21111) spin systems. Tetrahedral hs d^6 molecules are Jahn–Teller unstable, but the orbital degeneracy is associated with the e-orbital set and $FeCl_4^{2-}$ is found to be regular tetrahedral. Intermediate-spin d^6 species of any coordination number are rare. Low-spin d^6 species are Jahn–Teller unstable for several of the geometries of Figure 11.5. The one known example, $M(CO)_4$ (M = Cr, Mo, or W), has a C_{2v} geometry, compatible with Jahn–Teller requirements. High-spin d^5 (11111) systems are Jahn–Teller stable at the tetrahedral geometry, and $MnCl_4^{2-}$ has this structure. Low-spin d^5 species are not known but are expected to be C_{2v}. High-spin d^4 (11110) should behave similarly to is d^6. Low-spin d^3, d^2, d^1, and d^0 are not predicted to have a static Jahn–Teller effect. $TiCl_4$ (d^0) and VCl_4 (d^1) are tetrahedral.

For five coordinate geometries we concentrate on a few electronic configurations of interest. First we note (Figure 5.10) that the e' vibration of the tbp connects to the spy geometry via a C_{2v} structure. Low-spin d^6 (22200) systems are clearly unstable (Figure 9.7) at the tbp geometry, and distortion via the Jahn–Teller active e' mode is predicted. Intermediate-spin d^6 (22110) is stable at the tbp, and a dynamic effect is predicted at the spy. The low-spin form corresponds to the lowest singlet (all electrons paired) and the high-spin form to the lowest triplet if the two highest-energy electrons have parallel spins. $Cr(CO)_5$ (ls d^6) is found to be distorted all the way along the C_{2v} coordinate to square pyramidal, and we discuss some implications of the stable is d^6 tbp geometry in Section 11.4. Low-spin d^7 systems are Jahn–Teller unstable at the tbp geometry. $Re(CO)_5$ and $Mn(CO)_5$ are square pyramids. Low-spin d^8 molecules are Jahn–Teller stable in both geometries. Examples of the tbp and spy geometries are found for $Fe(CO)_5$ and $Ni(CN)_5^{3-}$, respectively.

For three-coordinate molecules, of which there are few simple examples since in order to obtain this coordination number rather bulky ligands often have to be used,[21,55] the Jahn–Teller active coordinate is of species e', which sends the trigonal plane to the T shape for E' states.

Using the level pattern of Figure 9.7 the ls d^6 and the hs d^8 and d^{10} species are Jahn–Teller stable at the trigonal planar geometry. The ls d^8 structure is unstable, and, in agreement with the prediction of the theorem, $Rh(PPh_3)_3^+$ is a T-shaped molecule.[224] However $Cr(CO)_3$ (ls d^6) and $Fe(CO)_3$ (hs d^8) are pyramidal molecules.[22] Neither species is Jahn–Teller unstable at the D_{3h} geometry ($^1A_1'$ for ls d^6, $^3A_2'$ for hs d^8) and the pyramidal structure may not be rationalized on the scheme. Thus for the first time distorted structures are found which have no immediate explanation using this approach.

The second-order effect is of rather restricted predictive use in these particular examples. The lowest-energy transition (Figure 9.7) for the ls d^6 and hs d^8 $M(CO)_3$ species at the D_{3h} geometry is $a_1' \rightarrow e'$, which leads to a transition density of species e' and the prediction of a T-shaped geometry. For both of these molecules this is not borne out; trigonal pyramids are observed. There is a higher-energy transition $e'' \rightarrow e'$ where the transition density contains a_2'', the correct symmetry species for pyramidalization. In the d^{10} species this "transition" is blocked and the molecule is predicted (and observed) to be planar. However we found for main group systems that it was the lowest-energy transition which was the geometry determining one. So, although there is an *a posteriori* explanation for the pyramidal geometry (the higher energy transition is more important here), the method is clearly not very good from the point of view of its general predictive power.

For six-coordinate molecules static Jahn–Teller effects are expected wherever the e_g orbital is unsymmetrically occupied as in hs d^4, is d^5, is d^6, ls d^7, ls d^8, and ls d^9. The is complexes are rare, and the ls d^8 system (22220) will be much more stable at the octahedral geometry as the hs form (22211) with the two e_g electrons having parallel spins. (We will however return to this case later.) This leaves the hs d^4, ls d^7, and ls d^9 systems. All three have E_g electronic states and the Jahn–Teller "active" coordinate is of species e_g. There is no bending vibration of this symmetry in this molecule, but there is an e_g stretching vibration of which either phase is allowed (Figure 11.7). The situation with two long and four

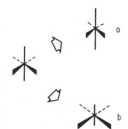

Figure 11.7 One component of the e_g stretching vibration of an octahedral MY_6 molecular showing two different phases leading to distorted molecules with (a) two long and four short bonds and (b) four long and two short bonds.

short bonds is the one most often found. The opposite occurs in a few examples. It is probably true to say that distortions of these molecules always give two long and two short bonds if (a) all the ligands are identical, and (b) the site containing the Jahn–Teller "active" ion in the solid is a regular octahedral one when a Jahn–Teller "inactive" ion occupies it. Thus in $Ba_2Zn_{1-x}Cu_xF_6$ ($x \leq 0.3$) the tetragonally compressed copper environment is similar to that of zinc which it replaces. We have commented earlier on one set of molecules where the distortion is only observable using X-ray crystallography when the sample is cooled. Table 11.3 shows a selection of octahedral systems which are distorted via an e_g stretching mode of the octahedron. Most of the examples are for the d^9 Cu(II) case where the distortions are often very dramatic.

Either the CFT or AOM may be used to derive an energy diagram for the distortion.[52] Group theory gives the following expression for the e_g stretching vibration:

$$\Delta r(e_g) = \frac{1}{\sqrt{12}} (2\Delta r_1 + 2\Delta r_2 - \Delta r_3 - \Delta r_4 - \Delta r_5 - \Delta r_6) \quad (11.1)$$

where $r_{1,2}$ and r_{3-6} are the axial and equatorial MY linkages, respectively. The CFT energy diagram for a D_{4h} $MY_4'Y_2''$ system is shown in Figure 11.8, for both $\rho = 1$ and 2 and also the AOM result for this species. If we now assume that $Dq' = Dq + \Delta Dq$ and $Dq'' = Dq - 2\Delta Dq$, where Dq is the octahedral value of this parameter and Dq' and Dq'' are the values for the ligands Y and Y', then the diagrams of Figures 11.9 and 11.10 result. The changes in Dq for axial and equatorial ligands are of the opposite sign since according to Equation 11.1 one set decreases and the other set increases its MY distances on distortion $O_h \rightarrow D_{4h}$. The equatorial ligands move twice as far, and thus there is a factor of 2 arising in Dq''. This is of course assuming that $\Delta(Dq) \propto -\Delta r$ by manipulation of Equation 8.7. Figure 11.8 shows an analogous result using the AOM, where $e_\lambda' = e_\lambda + \Delta e_\lambda$ and $e_\lambda'' = e_\lambda - 2\Delta e_\lambda$ for $\lambda = \sigma, \pi$. Note that within this approximation z^2 drops in energy by the same amount x^2-y^2 is raised in energy. So with equal occupation of these two orbitals (e.g., hs d^8) the system is stable at the octahedral geometry, but with unequal occupation (e.g., ls d^8, d^9) a distortion is energetically allowed. We still have no indication however which distortion (four long or four short bonds) will occur. The three orbitals derived from the t_{2g} set however split apart asymmetrically. Whether xz, yz lie to lower energy than xy depends on the sign of Δe_π or ΔDq. This will in general be positive for the case of four short and two long bonds (model a, Figure 11.7) and

Table 11.3 Some Jahn–Teller Distorted Molecules

Compound	Short Distances (Å)	Long Distances (Å)
CuF_2	4F at 1.93	2F at 2.27
$K_2CuF_4{}^a$	2F at 2.22	4F at 1.92
$KCuF_3$	2F at 1.96	4F at 2.07
$CuF_2 \cdot 2H_2O$	2F at 1.89	2F at 2.47
	$2H_2O$ at 1.93	
$CuCl_2$	4Cl at 2.30	2Cl at 2.95
$CsCuCl_3$	4Cl at 2.30	2Cl at 2.65
$CuCl_2 \cdot 2H_2O$	2Cl at 2.28	2Cl at 2.95
	$2H_2O$ at 1.93	
$CuCl_2 \cdot 2C_5H_5N$	2N at 2.02	2Cl at 3.05
	2Cl at 2.28	
$CuBr_2$	4Br at 2.40	2Br at 3.18
$\alpha\text{-}Cu(NH_3)_2Br_2$	$2NH_3$ at 1.93	2Br at 3.08
	2Br at 2.54	
$\beta\text{-}Cu(NH_3)_2Br_2$	$2NH_3$ at 2.03	4Br at 2.88
$Cu(DMG)_2$	4N at 1.94	2O at 2.43
$Cu(NH_3)_2Cl_2$	$2NH_3$ at 1.95	4Cl at 2.76
$Cu(NH_3)_4SO_4 \cdot H_2O$	$4NH_3$ at 2.15	$1H_2O$ at 2.59
		$1H_2O$ at 3.37
$Cu(NH_3)_6{}^{2+}$	$4NH_3$ at 2.07	$2NH_3$ at 2.62
CrF_2	4F at 2.00	2F at 2.43
$KCrF_3$	2F at 2.00	4F at 2.14
MnF_3	2F at 1.79	2F at 1.91
		2F at 2.09
$\gamma\text{-}MnO(OH)$	4O at 1.88	2O at 2.30
$\beta Mn(acac)_3$	2O at 1.948	4O at 2.000
$\gamma Mn(acac)_3$	4O at 1.929	2O at 2.116
$Mn(tropolonato)_3$	4O at 2.127	2O at 1.941

^{a} In earlier tabulations this species is quoted as having two short (1.95 Å) and four long (2.08 Å) bonds. A recent investigation[85] of the structure shows a superstructure with a tetragonally *elongated* Cu(II) environment.

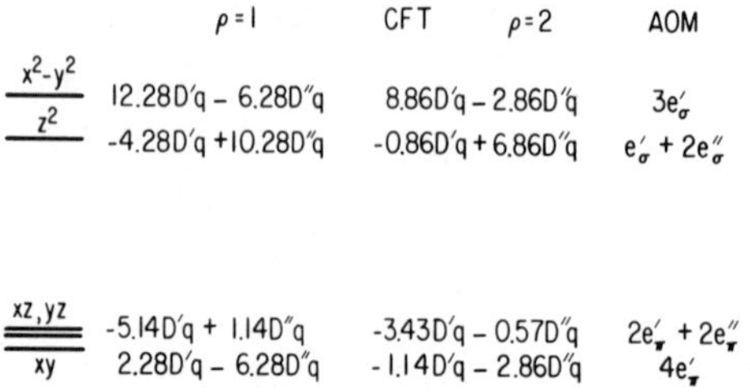

	$\rho=1$	CFT	$\rho=2$	AOM
$\dfrac{x^2-y^2}{z^2}$	$12.28D'q - 6.28D''q$ $-4.28D'q + 10.28D''q$		$8.86D'q - 2.86D''q$ $-0.86D'q + 6.86D''q$	$3e'_\sigma$ $e'_\sigma + 2e''_\sigma$

	$\rho=1$	$\rho=2$	AOM
$\dfrac{xz,yz}{xy}$	$-5.14D'q + 1.14D''q$ $2.28D'q - 6.28D''q$	$-3.43D'q - 0.57D''q$ $-1.14D'q - 2.86D''q$	$2e'_\pi + 2e''_\pi$ $4e'_\pi$

Figure 11.8 CFT and AOM splitting patterns for a $D_{4h}MY_4'Y_2''$ species. (In terms of spectroscopic crystal field parameters the energies are: z^2, $6Dq - 2Ds - 6Dt$; x^2-y^2, $6Dq + 2Ds - Dt$; xz,yz, $- 4Dq - Ds + 4Dt$; xy, $-4Dq + 2Ds - Dt$.)

negative for the reverse (model b). Table 11.4 gives the stabilization energies associated with each model for several configurations of interest. For the FeF_2 species (which exhibits a distorted rutile structure in the solid) the observed direction of the distortion is in agreement with the theory ($t_{2g}^4 e_g^2$). CoF_2 however should exhibit the opposite distortion ($t_{2g}^5 e_g^2$), but none is found experimentally.

The general observation of four short and two long bonds (model a in Figure 11.7) in these octahedrally based molecules with an orbital degeneracy involving the e_g pair may be understood by looking at the effect of higher energy orbitals, specifically the $(n + 1)s$ orbital, on the energies of the d orbitals. In the distorted octahedral geometry of D_{4h} point symmetry both z^2 and $(n + 1)s$ orbitals transform as a_{1g}. Thus, as in the square planar structure discussed in Section 9.3, z^2 will be stabilized by such ds mixing. Clearly the axial elongation route will be energetically preferred as a result since two electrons occupy z^2 for a d^9 species in this geometry compared to the presence of a single z^2 electron for the axially compressed alternative. Viewed in this light the distorted Cu(II) environments might then be better regarded as second-order (rather than first-order) Jahn–Teller distortions.

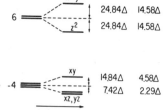

Figure 11.9 CFT splitting pattern for an e_g distortion of an octahedral MY_6 molecule (units are Dq.) ΔDq is positive for case (a) (Figure 11.5) and negative for case (b).

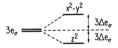

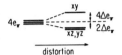

Figure 11.10 AOM splitting pattern for an e_g distortion of the octahedral molecule.

Table 11.4 Stabilization Energies on Distortion of the Octahedron for four short and two long (Model a) and four long and two short (Model b) bonds[a]

Electronic Configuration	Model a	Model b
d^9 22221	$3\Delta e_\sigma$, $(14.58\Delta Dq)$, $24.84\Delta Dq$	$3\Delta e_\sigma$, $(14.58\Delta Dq)$, $24.84\Delta Dq$
ls d^7 22210	$3\Delta e_\sigma$, $(14.58\Delta Dq)$, $24.84\Delta Dq$	$3\Delta e_\sigma$, $(14.58\Delta Dq)$, $24.84\Delta Dq$
hs d^6 21111	$2\Delta e_\pi$, $(2.29\Delta Dq)$, $7.42\Delta Dq$	$4\Delta e_\pi$, $(4.58\Delta Dq)$, $14.84\Delta Dq$
hs d^7 22111	$4\Delta e_\pi$, $(4.58\Delta Dq)$, $14.84\Delta Dq$	$2\Delta e_\pi$, $(2.29\Delta Dq)$, $7.42\Delta Dq$

[a] The three entries are for the AOM and CFT ($\rho = 1$) models (values for $\rho = 2$ are in parentheses).

The idea that asymmetric occupation of degenerate orbitals is often an energetically unfavorable situation enables symmetrical structures to be excluded as candidates for the geometries of molecules with certain electronic configurations, but it is difficult *a priori* to predict the stable geometry. In the next section we show how this problem may be overcome by calculating the d-orbital contribution to the energy for various geometries and electron configurations.

11.3 CFSE or MOSE Arguments

On a d orbital-only model the CFSE or MOSE for various geometric alternatives as a function of d-orbital configuration may readily be computed to see which geometry is preferred. The result of Figure 11.11 is a familiar one where the energy difference between octahedral and tetrahedral coordination is plotted. A plot of this type is able to cast light on the observation that tetrahedral d^8 and d^3 systems are rare but many other configurations are represented by a plentiful number of examples.[81]

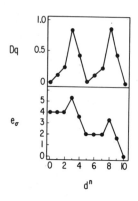

Figure 11.11 Difference in (*a* CFSE and (*b*) MOSE between octahedral and tetrahedral MY_6 and MY_4 complexes as a function of d^n for high-spin systems.

We need to remember however in using plots such as these that the CFSE or MOSE represents only a part of the stabilization energy of the complex. In this section instead of looking at the energy differences between structures with different coordination numbers we look at geometric alternatives of systems with the same coordination number.

Rather than calculate Walsh diagrams for each distortion coordinate we compute stabilization energies for a limited number of extreme geometries. Table 11.5 shows MOSEs and CFSEs for three different four-coordinate geometries using a σ-only model (e_σ and f_σ) for the molecular orbital approach. (This is a simplification we can make by knowing that $e_\pi/e_\sigma < 1$ and often ~0.25 or less.) Two sets of results are given for the crystal field approach for $\rho = 1$ and 2. First we examine the molecular orbital results. In every case where the tetrahedral molecule was predicted to be Jahn–Teller unstable, or alternatively where there is asymmetric occupation of degenerate orbitals, the d-orbital stabilization energy is larger for some other geometry. On d-orbital grounds the molecule should distort. But what fixes the actual size of the distortion away from the tetrahedral structure? In Chapter 10 it was emphasized that ligand–metal d orbital interaction was only one of the forces involved in MY interaction. Interactions with metal s and p orbitals were probably more important in numerical terms. If the d orbitals, and the electrons in them, are ignored then the geometry expected for a MY_n complex would be given simply by the VSEPR scheme (i.e., trigonal planar for MY_3, tetrahedral for MY_4, etc.). In these cases the VSEPR geometries are also those expected on the basis of minimum ligand–ligand nonbonded repulsions. We call these combined forces ligand–ligand terms. If the VSEPR structure is either most stable on d-orbital grounds or if the differential MOSE is small or zero for some other geometry, then the observation of the VSEPR structure is expected. If there is a geometry where the d-orbital stabilization

Table 11.5 MOSE and CFSE for Some Four-Coordinate Geometries

Electronic Configuration	MOSE[a]			CFSE[b]		
	Octahedral Cis Divacant	Square Planar	Tetrahedral	Octahedral Cis Divacant	Square Planar	Tetrahedral
ls d^5 22100	8.0	8.0	6.667	−13.72 (−14.57)	−14.86 (−24.84)	−8.90
is d^5 21110	6.5	7.0	5.333	−9.86 (−13.28)	−12.29 (−17.54)	−4.45
hs d^5 11111	4.0	4.0	4.0	0 (0)	0 (0)	0
ls d^6 22200	8.0[c]	8.0[c]	5.333	−16.0 (−16.0)	−16.0 (−29.12)	−7.12
is d^6 22110	6.5	7.0	5.333	−12.15 (−14.71)	−15.72 (−22.56)	−7.12
hs d^6 21111	4.0	4.0	4.0	−2.29 (−5.14)	−3.43 (−5.14)	−2.67
ls d^7 22210	6.5	7.0	4.0	−14.43 (−16.14)	−16.86 (−26.84)	−5.34
hs d^7 22111	4.0	4.0	4.0	−5.715 (−6.57)	−6.86 (−10.28)	−5.34
ls d^8 22220	5.0	6.0	2.667	−12.86 (−16.28)	−17.72 (−24.56)	−3.56
hs d^8 22211	4.0[c]	4.0[c]	2.667	−8.0 (−8.0)	−8.0 (−14.56)	−3.56
d^9 22221	2.5	3.0	1.333	−6.43 (−8.14)	−8.86 (−12.28)	−1.78
d^{10} 22222	0	0	0	0 (0)	0 (0)	0

[a] Units of e_σ (σ-only model).

[b] Units of Dq, values for $\rho = 1$ (values for $\rho = 2$ in parentheses).

[c] In these cases the MOSE for two nontetrahedral geometries are equal. In order to resolve the structural problem (also for the three coordinate results of Table 11.6) the f_σ term needs to be considered. The general result is that the structure with the largest number of cis ligands is most stable, namely the octahedral cis divacant structure or the fac trivacant three coordinate geometry. This result is derived from the orbital sharing rule of Chapter 6.

187

is significantly larger than for the VSEPR structure, then the molecule will distort. The extent of the distortion should be related to the size of the differential stabilization energy, and this is indeed the case experimentally. Thus ls d^6 Cr(CO)$_4$ has a larger MOSE at the cis divacant geometry relative to tetrahedral than hs d^8 Fe(CO)$_4$. The former molecule is much closer to this geometry than the iron analog (Figure 11.1). Low-spin d^8 molecules have the largest differential stabilization from tetrahedral at the square planar geometry and are always found in this arrangement. d^9 systems have half the ls d^8 energy difference and as noted before are often found with D_{2d} and sometimes D_{4h} geometries. ([Cu(adeninium)$_2$Br$_2$]Br$_2$ actually contains[138] a CuY$_4^{2+}$ ion where the bond angles 144.7(2)° are exactly halfway between those of tetrahedral (109.5°) and square planar (180°) geometries.) In solution spectral evidence suggests that the structures lie somewhere between tetrahedral and square planar extremes. Clearly this is a method which allows simple calculation of the "distorting force" away from the tetrahedral geometry.

Because of the σ-only nature of the model, any orbital degeneracy involving the e orbitals does not show up in the numbers of Table 11.5 as a destabilization of the tetrahedral structure. However in most of our examples the lowest-energy d orbitals (those involved in π interactions) are fully occupied for nontetrahedral structures. Since for π donors these orbitals are M–Y antibonding there is no net π bonding in these geometries. For π acceptors, π bonding is at a maximum here. At the tetrahedral structure (Figure 9.6) the higher-energy t_2 orbital set are involved in π interactions. Thus systems with π donor ligands might be expected to resist distortion away from tetrahedral and systems with π acceptors the opposite. We noted earlier that high-spin d^8 Fe(CO)$_4$ was distorted to C_{2v} but the isoelectronic NiX$_4^{2-}$ species appeared undistorted.

Some rather different and less satisfactory results are obtained if the CFSE results of Table 11.6 are used instead. Whereas the hs d^6 (21111) molecule had no driving force away from tetrahedral on the σ-only mo-

Table 11.6 MOSE for Some Three-Coordinate Structures (Units e_σ)

Electronic Configuration	Octahedral Fac Trivacant	Trigonal Plane	T shape
ls d^6 22200	6.0	4.5	6.0
ls d^8 22220	3.0	2.25	4.73
hs d^8 22211	3.0	2.25	3.0
d^9 22221	1.5	1.125	2.37
d^{10} 22222	0	0	0

lecular orbital model (and $FeCl_4^{2-}$ is tetrahedral), the CFT predicts a distorted structure for both $\rho = 1$ and 2, although the difference in CFSE between tetrahedral and cis divacant geometries is smaller than for the ls d^6 system where a distorted structure is found experimentally. More importantly the CFT never predicts a cis divacant geometry. For ls d^6, hs d^8 with $\rho = 1$, the cis divacant and square planar structures are of equal energy. (For $\rho = 2$ the D_{4h} geometry is favored.)

Three-coordinate molecules were an example where neither first- or second-order Jahn–Teller results predicted the observed geometries for ls d^6 and hs d^8 configurations. Table 11.6 shows MOSEs for various three-coordinate geometries. The pyramidal structure is clearly predicted for both species with the larger distortion away from trigonal planar (Figure 11.1) of $Cr(CO)_3$ reflected in the larger differential MOSE for this electronic configuration. Also note that the $Cr(CO)_3$ structure is about two thirds the way to the d-orbital-determined geometry (differential stabilization energy = $1.5e_\sigma$). $Cr(CO)_4$ is very close (differential stabilization energy = $2.25e_\sigma$) and $Ni(CN)_4^{2-}$ (differential stabilization energy = $3.3e_\sigma$) adopted exactly the d orbital-preferred structure.

Five-coordinate molecules are approachable along the same lines. Low-spin d^8 species are predicted to be square pyramidal, but with a small driving force away from the tbp geometry on both AOM and CFT models (Table 11.7). $Fe(CO)_5$ is found as the tbp structure but with a low barrier to rearrangement, probably via the square pyramidal geometry and a Berry pseudorotation (Figure 6.13). $Ni(CN)_5^{3-}$ exists in the crystal in two different geometries—as a square pyramid and as a trigonal bipyramid which is distorted toward the spy structure. (See Figure 3.3 for

Table 11.7 MOSE and CFSE for Some Five-Coordinate Geometries

Electronic Configuration	MOSE[a] tbp (D_{3h})	spy (C_{4v})	C_{2v}	CFSE[b,c] tbp (D_{3h})	spy (C_{4v})
ls d^6 22200	7.75	10	8.75	-12.49 (-12.50)	-20.0 (-20.0)
is d^6 22110	7.75	8	7.75	-12.49 (-12.50)	-14.86 (-18.28)
ls d^7 22210	6.625	8	7.0	-12.46 (-13.32)	-17.43 (-19.14)
ls d^8 22220	5.5	6	5.5	-12.42 (-14.14)	-14.86 (-18.28)
d^9 22221	2.75	3	2.75	-6.21 (-7.07)	-7.43 (-9.14)
d^{10} 22222	0	0	0	0	0

[a] Units of e_σ (σ-only model).
[b] Units of Dq.
[c] Values for $\rho = 1$ (values for $\rho = 2$ in parentheses).

a pictorial view of the spectrum of five-coordinate geometries.) On applying pressure the reversible transformation tbp $\rightarrow$ spy occurs. $Cr(CO)_5$ $Mo(CO)_5$, $W(CO)_5$ (ls d^6), $Mn(CO)_5$, $Re(CO)_5$, and $Co(CN)_5^{3-}$ (ls d^7) are square pyramids commensurate with the larger driving forces from the tbp structure for these electronic configurations. An alternative geometry which relieves the Jahn–Teller instability of the tbp for ls d^6, d^7 config- urations is the C_{2v} one (Figure 11.12) reached via the opposite phase of the tbp distortion which leads to the square pyramid. In none of the sys- tems we have mentioned however is it the lowest-energy structure, but as we see later it may be important in thermal rearrangements of these species.

This direct energy evaluation method using the molecular orbital ap- proach then gives a good idea as to the geometry a molecule should have. Combined with the knowledge of structural instability of asymmetric oc- cupation of degenerate orbitals it is a useful way to understand transition metal stereochemistry. We have used both CFT and AOM to quantita- tively estimate energy changes. The correlation between theory and ex- periment showed that the CFT could usually give similar results to the AOM by a judicious choice of the parameter ρ. But whereas chemical intuition gives a feeling for the relative importance of σ and π bonding effects on the MO model, the choice of the correct ρ value (if in fact there is just a single value applicable to all systems) is not obvious. For several systems, especially those with π acceptor ligands ML_n, π bonding will be important. This is discussed in detail in Section 12.2 and is an important structural feature which CFT is unable to consider.

Some geometries contain an angular degree of freedom, and the AOM may be used to construct a Walsh diagram describing the energy changes within this coordinate. Figure 9.8 shows such a diagram for the variation of the axial-basal angle in C_{4v}, spy MY_5 systems. Occupancy of the a_1 (z^2) or b_1 (x^2-y^2) orbitals will encourage bending, and the geometric re- sults shown in Table 11.8 support this idea. Of some interest is the ge- ometry change with spin state of the hemoglobin species.[167] It has been argued that the reason for the observation of an out-of-plane hs Fe(II) atom in hemoglobin but in-plane ls Fe(II) atom in oxyhemoglobin was the difference in ionic radius. That for hs Fe(II) was suggested to be too large to allow the atom to fit into the porphyrin ring (Figure 11.13). The radius of the low-spin atom was suggested to be smaller, allowing it to drop into

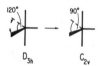

Figure 11.12   The C_{2v} structure (equatorial angles 90, 135, and 135°) derived from the tbp.

Table 11.8 Some Bond Angles in Square Pyramidal Units as a Function of Electronic Configuration

Electronic Configuration	Compound	Apical–M–Basal Angle (degrees)
d^{10} 22222	Cu^I macro $\cdot$ CO^a	117
hs d^8 22211	Ni(5-ClSalenNEt$_2$)$_2$	100.4
	[Ni(dmp)Cl$_2$]$_2$ $\cdot$ 2CHCl$_3$	100.9
	Ni(bddae)(NCS)$_2$	99.9
	[Ni(tpen)](ClO$_4$)$_2$ $\cdot$ MeNO$_2$	98.8
ls d^8 22220	Ni(bda)Br$_2$	93.9
	Ni(DSP)I$_2$	92.4
	Ni(CN)$_5{}^{3-}$	101.0
	Co(CNC$_6$H$_5$)$_5{}^{2+}$	101.8
	Co(CN)$_5{}^{3-}$	97.6
hs d^6 21111	Deoxyhemoglobin (FeII)	110
ls d^6 22200	Oxyhemoglobin (FeII)	~90
hs d^5 11111	Chlorohemin (FeIII)	93
ls d^5 22100	Cyanomethemoglobin (FeIII)	90–92

^{a} Macro = difluoro-3,3'-(trimethylenedinitrito)bis(2-butanone oxamate).

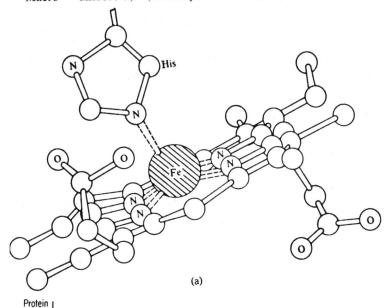

(a)

(b)

Figure 11.13 (a) Geometric arrangement in hemoglobin Reproduced with permission from J. E. Huheey, *Inorganic Chemistry*, 2nd ed., Harper and Row (1978). (b) Mechanical picture showing how the Fe(II) atom sits above the plane in the hs and in the plane in the ls forms.

the plane. However Sn(IV) and Mo(IV) species with "radii" larger than that for hs Fe(II) also form porphyrin complexes with in-plane metal atoms, which suggests that another explanation of the effect is in order, and this the molecular orbital approach provides.

11.4 Excited-State Geometries

The geometries of excited electronic states are as equally accessible on the AOM approach as the ground-state geometry. (There is always the proviso in such arguments that a given electronic state is well described by a single electronic configuration and that configuration interaction is not important.) In the tables associated with the previous section we calculated the equilibrium geometry as a function of d-electron configuration, which is intimately connected with the nature of the electronic state. Thus the ground-state geometry of a ls d^8 system (22220) is square planar. Promotion of an electron from the HOMO to the LUMO leads to the configuration 22211, an identical set of numbers to the high-spin (triplet) d^8 case. The geometry of the first excited singlet state should then be similar to that of the lowest triplet, namely, C_{2v}. In fact there is good evidence for such a geometry. On photochemical excitation of trans PtY_2Y_2' species the cis species is produced via a nondissociative route. One simple way of proceeding is through the C_{2v} distorted tetrahedron[9] (Figure 11.14). Similarly $W(CO)_4CS$ (ls d^6) on visible excitation in a low-temperature matrix exchanges axial and basal ligands[173] probably via a trigonal bipyramidal excited state (22110) as in Figure 11.15. Table 11.7 shows that this configuration has a very small driving force away from

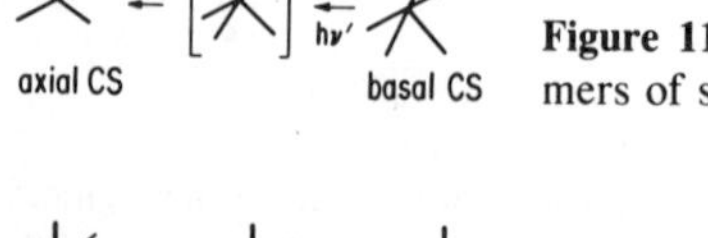

Figure 11.14 Photochemical interversion of square planar $Pt^{II}Y_2Y_2'$ species.

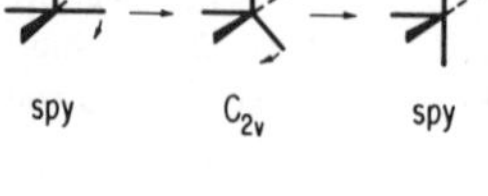

Figure 11.15 Photochemical interconversion of the two isomers of square pyramidal $W(CO)_4CS$.

Figure 11.16 Alternative route for apical–basal ligand exchange in square pyramidal molecules.

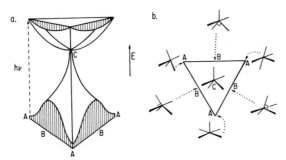

Figure 11.17 (a) Schematic potential energy surface for interconversion of square pyramidal d^6 M(CO)$_5$ molecules via C_{2v} and trigonal bipyramidal (D_{3h}) structures for the ground (22200) and first excited singlet (22110) electronic states. (b) Plan of the three-dimensional figure in (a). Structures A and B are spy and C_{2v} geometries respectively, and C is the trigonal bipyramid.

the tbp, smaller than that for Fe(CO)$_5$ which experimentally is found as a tbp. Thermally however there is another route for axial/basal ligand exchange which is probably of lower energy (see Table 11.7), via a C_{2v} geometry (Figure 11.16). A potential energy surface for thermal and photochemical rearrangement within this system[26] is shown in Figure 11.17 in schematic form.[89]

11.5 Quantitative Calculations

Table 11.9 shows the results of two sets of EHMO calculations[24,54] to determine the angular geometry of M(CO)$_n$ species. The only differences between the two are in parametrization of the molecular orbital variables (see Section 2.4) of the method. The same bond lengths were used irrespective of coordination number or geometry. The agreement with experiment is generally very good. (We include some noncarbonyl structures as examples, which is not strictly valid since the calculations were performed on systems with CO ligands.) While the D_{2d} geometry is predicted for a d^9 M(CO)$_4$ species by both sets of calculations, the C_{3v} structure is also calculated to be close in energy above it. Both are observed for Co(CO)$_4$. Analogous molecular orbital arguments lead to the prediction of a C_{2v} structure for hs d^8 Fe(CO)$_4$ but with a close-lying C_{3v} or C_s structure. The molecule rearranges[45] thermally via a route which is the permutational opposite of that found for SF$_4$ (Figure 6.12) and is consistent with a transition state of this symmetry. In these low-symmetry situations there is no substitute for numerical calculations of the energy differences. The CFT or AOM, being restricted to d orbital–ligand interactions and

Table 11.9 Calculated Geometries of Binary Transition Metal Carbonyls Using the EHMO Method

d-Orbital Population	I (Ref. 24)	II (Ref. 54)	Examples[a]
Coordination number 3			
d^5 22100	C_{3v} ($\theta = 63°$)	C_{3v} ($\theta = 58°$)	$V(CO)_3$ C_{3v} or D_{3h}
d^6 22200	C_{3v} ($\theta = 60°$)	D_{3v} ($\theta = 57°$)	$Mo(CO)_3$ C_{3v} ($\theta = 65°$)
22110	C_{2v} ($\alpha = 170°$)		
21111	D_{3h}		
d^7 22210	C_{2v} ($\alpha = 170°$) ($\theta = 90°$)	C_{3v} ($\theta = 63°$)	
22111	D_{3h}		
d^8 22220	C_{2v} ($\alpha = 175°$) ($\theta = 90°$)	C_{2v} ($\alpha = 172°$) ($\theta = 90°$)	$Rh(PPh_3)_3{}^+$ ($\theta = 90°$, $\alpha = 159°$)
22211	C_{3v} ($\theta = 73°$)		$Fe(CO)_3$ C_{3v} ($\theta = 72 \pm 3°$)
d^9 22221	C_{2v} ($\alpha = 150°$)	C_{3v} ($\theta = 75°$)	$Co(CO)_3$ C_{3v} or D_{3h}
d^{10} 22222	D_{3h}	C_{3v} ($\theta = 82°$)	$M(L_2)_3$, M = Ni, Pd, Pt; L_2 = CO, N_2 D_{3h}
Coordination number 4			
d^5 22100	C_{2v} (100°, 160°)	C_{2v} (100°, 160°)	$V(CO)_4$ T_d or D_{4h}?
d^6 22200	C_{2v} (90°, 170°)	C_{2v} (95°, 165°)	$Mo(CO)_4$ D_{2v} (107°, 174°)
22110	D_{4h}		
21111	T_d		
d^7 22210	D_{4h}	C_{2v} (135°, 150°)	
22111	T_d		
d^8 22220	D_{4h}	D_{2d} (150°)	$Ni(CN)_4^{2-}$ D_{4h}
22211	C_{2v} (110°, 135°)		$Fe(CO)_4$ C_{2v} (~120°, ~145°)
d^9 22221	D_{2d} (132°)	D_{2d} (135°)	$Fe(CO)_4^-$, $Co(CO)_4$ C_{3v} ($\theta \sim$ 100°) also $Co(CO)_4$ D_{2d}, $Rh(N_2)$ D_{2d}
d^{10} 22222	T_d	T_d	$M(L_2)_4$ M = Ni, Pd, Pt; L_2 = CO; M = Ni; L_2 = N_2 T_d

Table 11.9 (Continued)

Coordination number 5

d^5	22100	D_{3h}	C_{4v} ($\theta = 94°$)	$V(CO)_5\ D_{3h}$?
d^6	22200	C_{4v} ($\theta = 93.5$)	C_{4v} ($\theta = 93°$)	$W(CO)_4CS$, $Cr(CO)_5\ C_{4v}$ ($\theta = 91–94°$)
	22110	D_{3h}		
	21111	D_{3h}		
d^7	22210	C_{4v} ($\theta = 98°$)	C_{4v} ($\theta = 98°$)	$Mn(CO)_5$, $Re(CO)_5\ C_{4v}$ $Cr(CO)_5^-$ ($\theta \sim 95°$)
	22111	D_{3h}		
d^8	22220	D_{3h}	D_{3h}	$Mn(CO)_5^-$, $Fe(CO)_5\ D_{3h}$
	22211	C_{2v}		

Angles

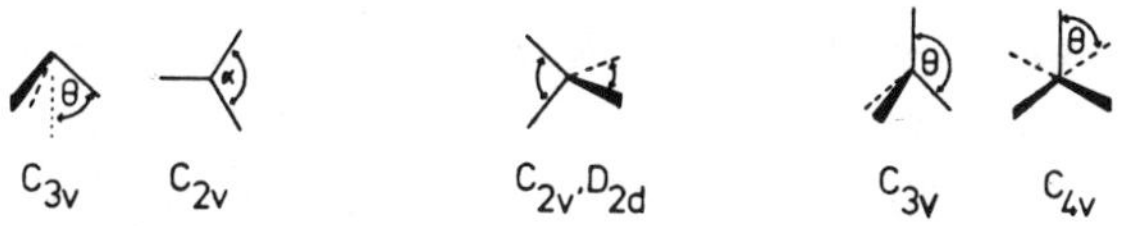

C_{3v} C_{2v} C_{2v},D_{2d} C_{3v} C_{4v}

---

[a] Only well-characterized molecules appear in this column. A question mark follows some structures which are based on information not as good as the rest. All the examples given have π acceptor ligands. A third numerical study on these systems is to be found in D. A. Pensak and R. J. McKinney, *Inorg. Chem.* **18** 3407 (1979).

not simply adapted to include the effects of *s*-, *p*-, and *d*-orbital mixing is too crude a model to resolve such differences. Similarly studies on larger molecules, for example systems containing organometallic units, require numerically based molecular orbital methods.[94,98–101,127,128,178]

12 Bond Lengths and σ and π Bonding

The simple angular overlap approach for main group systems gave a ready rationalization of the relative lengths of the nonequivalent MY linkages in species such as IF_7, BrF_5, SF_4 and ClF_3. It also provides a way to understand bond length variations in transition metal systems. In this chapter we look at the relative bond lengths expected on σ and π bonding grounds in a simplistic fashion. In the next chapter we look at the molecular orbital structure of transition metal complexes in greater detail.

12.1 σ Bonding

One very general observation is that as the coordination number of transition metal increases the MY bond lengths tend to increase. Put a different way, the "radius" of the metal atom is smaller in four than in six-coordinate systems, for example.[187] Supporting evidence for bond weakening comes from studies of vibrational frequencies (or more strictly force constants). MY stretching force constants decrease with increasing coordination number. For example the totally symmetrical stretching vibrations of $HgCl_2$, $HgCl_3^-$, and $HgCl_4^{2-}$ occur at 360, 293, and 270 cm^{-1}, respectively.[1] In Chapter 10 the importance of metal s, p interactions with the ligands was stressed. Why the average bond stabilization energy per ligand in general decreases with increasing coordination is readily seen by using a very simple model. Figure 12.1 shows diagrams for ligand σ–metal s, p interactions, in terms of the e_σ and f_σ parameters and the number of ligands, for the tetrahedral and octahedral coordination geometries. With two electrons in each bonding orbital, the total stabilization energy per MY linkage is

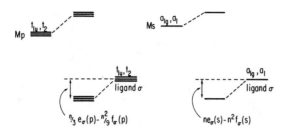

Figure 12.1 AOM diagrams for metal p and s–ligand σ interaction in octahedral and tetrahedral MY_n structures.

$$\sum_i (\sigma) = 2\left\{ e_\sigma(s) + e_\sigma(p) - n\left[f_\sigma(s) + \frac{1}{3}f_\sigma(p) \right]\right\} \qquad (12.1)$$

As the number of ligands n increases, the bond stabilization energy *per bond* falls. Clearly this electronic effect may be only one contribution. Increasing steric crowding as the coordination number increases for example will also lead to increasing MY bond lengths as discussed in Section 7.3.

We may use a similar method to compare bond lengths between n-coordinate systems with different geometries. For example d^8 Ni(II) forms a series of four-coordinate, approximately tetrahedral systems (high spin) and another series of square planar (low-spin) complexes. In the exactly tetrahedral structure the stabilization energy per bond is $\frac{2}{3}e_\sigma$. In the square planar arrangement $\Sigma_i(\sigma) = \frac{3}{2}e_\sigma$. Thus invariably shorter bonds to nickel are found in square planar compared to tetrahedral geometries. Typical bond lengths are Ni–N (e.g., py, bipy): 1.86 and 1.96 Å; Ni–P: 2.14 and 2.28 Å; Ni–S: 2.15 and 2.28 Å; Ni–Br: 2.30 and 2.36 Å, square planar results first. For d^7 Co(II), $\Sigma_i(\sigma)$ is simply e_σ and $\frac{7}{4}e_\sigma$ in tetrahedral (hs) and square planar (ls) environments, respectively. Typical bond lengths are Co–O: 1.86 and 2.06 Å; Co–N: 1.85 and 2.01 Å, square planar results first. In this particular case we do not have to resort to consideration of the terms in f_σ.

Since the metal d orbitals are oriented in specific directions, changes in particular bond lengths ought to be seen when electrons are added to these (often σ and π) MY antibonding orbitals as the d-electron configuration is changed. One striking example of this effect was shown in the seven-coordinate molecule of Figures 10.8 and 10.9; and of course the double-humped behavior of the M–O distance in octahedrally coordinated

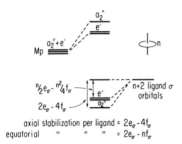

Figure 12.2 General method of estimating axial and equatorial bond stabilization energies for bipyramidal AY_{n+2} molecules using the AOM. Metal p–ligand σ interaction is shown here and $n = 3, 4, 5$, etc. An analogous method applies to d-orbital interactions where for $n = 3, 5, 6$, etc. the d orbitals split into $a_1(z^2)$ and $e(xy, x^2-y^2)$.

metal ions as a function of d^n of Figure 10.4d has a similar explanation. In order to be able to trace the effect of d-electronic configuration on bond lengths we need first to have a basis for understanding relative bond lengths in d^0 species. These are accessible by employing the same molecular orbital method used for main group systems in Chapter 6. The Ms–$Y\sigma$ interaction is isotropic and so is independent of whether the bonds are axial or equatorial (e.g., in a tbp) or apical or basal (e.g., in a spy). But interactions with p and d orbitals do depend upon the geometry. Figure 12.2 shows the method[27] of calculating the stabilization energies associated with axial and equatorial MY bonds in the bipyramidal series. p_z interacts only with the axial ligands and $p_{x,y}$ with the equatorial ones. The stabilization energy per axial MY linkage is clearly $2e_\sigma - 4f_\sigma$ and per equatorial linkage, $2e_\sigma - nf_\sigma$. The result is shown graphically in Figure 12.3, and it implies that on a p-orbital model $r_{ax}/r_{eq} > 1$ for the tbp, but $r_{ax}/r_{eq} < 1$ for the MY_7 pentagonal bipyramid (pbp) and MY_8 hexagonal bipyramid (hbp) structures as discussed in Chapter 6. The corresponding results for d-orbital interactions are obtained in a similar fashion. In this case to calculate the bond stabilization energies the quartic term (containing f_σ) needs to be divided into axial and equatorial contributions. In general we may write for the interaction energy of each symmetry type

$$\epsilon = (m_1 + m_2)e_\sigma - (m_1 + m_2)^2 f_\sigma \qquad (12.2)$$

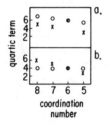

Figure 12.3 Quartic contributions to axial and equatorial bond stabilization energies via (*a*) central-atom p–ligand σ interaction and (*b*) central-atom d–ligand σ interaction. These plots represent the formulae $4 + n/2$ (axial, *d*), $1 + (11n/6)$ (equatorial, *d*), 4 (axial, *p*), and n (equatorial, *p*). Circles, axial linkages; crosses, equatorial linkages.

where m_1 and m_2 are the contributions from the two sets of symmetry-unrelated orbitals. Then the stabilization energy contribution to set 1 may be written as $m_1 e_\sigma - (m_1^2 + m_1 m_2) f_\sigma$ and that of set 2 as $m_2 e_\sigma - (m_2^2 + m_1 m_2) f_\sigma$, achieved by splitting the cross term appearing in the quartic contribution of Equation 12.2 equally between the two ligand sets. Using this method we find

$$\Sigma_{ax}(\sigma) = 2e_\sigma - \left(2 + \frac{n}{2}\right) f_\sigma$$

$$\Sigma_{eq}(\sigma) = 2e_\sigma - \frac{27n}{16} f_\sigma$$

(12.3)

a result plotted out in Figure 12.3. On the d orbital-only model, $r_{ax}/r_{eq} > 1$ for all these species (except the octahedron). The results of some Extended Hückel calculations[27] on these systems are shown in Figure 12.4 for the case of σ ligands only. The results indicate (for a particular set of parameters for s, p, and d orbitals) that whereas the relative bond lengths are set by Mp–$Y\sigma$ interactions, the Md–$Y\sigma$ interactions contribute significantly to the overall bond overlap populations for these d^0 systems. The detailed results are in broad agreement with our perturbation approach. For the spy geometry we showed in Chapter 6 that $r_{ap}/r_{ba} < 1$ on the p orbital-only model. Including d orbitals leads to the same result and specifically $\Sigma_{ap}(\sigma) = 2e_\sigma - 4f_\sigma$, $\Sigma_{ba}(\sigma) = 2e_\sigma - 1\frac{1}{2}f_\sigma$. Table 12.1

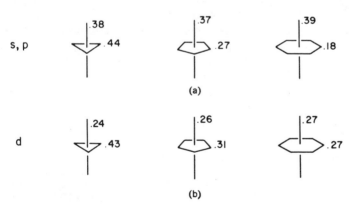

Figure 12.4 Calculated MY-bond overlap populations using the EHMO method for (*a*) metal p and s–ligand σ and (*b*) metal d–ligand σ interactions. (In a calculation where metal p and d orbitals were both included, the resultant values of the bond overlap populations were slightly different from the sum of the corresponding values in (*a*) and (*b*). Thus the "ligand additivity" is not quite exact.) From Ref. 27.

Table 12.1 Bond Lengths in Some
Square Pyramidal d^0 Species[a]

	Apical	Basal
$Nb(NMe_2)_5$	1.98	2.04
$InCl_5^{2-}$	2.42	2.46
$Sb(C_6H_5)_5$	2.12	2.22

[a] The latter two examples might also be regarded
as d^{10} species.

shows some structural results for d^0 systems which bear out the theoretical conclusions of this section.

Adding electrons to the system may result in occupation of MYσ antibonding orbitals and differential bond weakening processes may occur. For the square pyramidal molecule, the ls d^6 configuration should have a similar r_{ap}/r_{ba} ratio to the d^0 system, as found for the example in Table 12.2. Many other species however show little difference in bond length.[157] However with electron(s) in the next orbital (z^2) longer apical than basal bond lengths are expected since this orbital is more strongly M–Y_{ap} than M–Y_{ba} antibonding (Figure 12.5). Thus for both $Co(CN)_5^{3-}$ (ls d^7, $(z^2)^1$) and $Ni(CN)_5^{3-}$ (ls d^8, $(z^2)^2$) the M–C_{ap} length is significantly larger than its basal counterpart. The higher-energy $x^2–y^2$ orbital is only involved in interactions with the basal ligands however. For the high-spin d^8 systems with one electron each in $x^2–y^2$ and z^2, there is no experimental difference in axial and equatorial bond lengths (Table 12.2) consistent with the location of the two highest-energy d electrons in orbitals which overall are equally antibonding between the metal and apical and basal ligands. Analogously for d^9 systems the axial MY bond length is much longer than the basal.

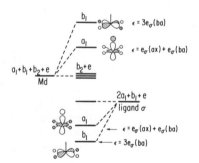

Figure 12.5 Details of the AOM d orbital-only diagram for a square pyramidal molecule showing that $x^2–y^2$ is only involved in metal–basal ligand bonding but z^2 is only in both basal and apical interactions. Occupation of the lowest- and highest energy a_1 orbitals then leads to a zero MY$_{ap}$ bond order.

Table 12.2 Some Representative Bond Lengths (Å) in Square Pyramidal Molecules

Electronic Configuration	Molecule	Apical MY Distance	Basal MY Distance	Bond
d^9 22221	$Cu(pyNO)_2(NO_3)_2$	2.44	1.96	Cu–O, Cu–N
	$Cu(dmg)_2$	2.30	1.95	Cu–O, Cu–N
hs d^8 22211	$Ni(5\text{-}ClSalenNEt_2)_2$	1.98	2.00	Ni–N
	$[Ni(dmp)Cl_2]$ $2CHCl_3$	2.06	2.07	Ni–N
	$Ni(bddae)(NCS)_2$	1.97	1.95	Ni–N
	$[Ni(tpen)](ClO_4)_2 \cdot MeNO_2$	2.10	2.10	Ni–N
ls d^8 22220	$Ni(bda)Br_2$	2.70	2.33	Ni–Br
	$Ni(CN)_5^{3-}$	2.17	1.85	Ni–C
	$Ni(DSP)I_2$	2.79	2.19	Ni–S
ls d^7 22210	$Co(CN)_5^{3-}$	2.01	1.90	Co–C
	$Co(CNC_6H_5)_5^{3-}$	1.95	1.84	Co–C
ls d^6 22200	$Ru(CO)(PPh_3)_2((CF_3)_2C_2S_2)$	2.27	2.35	Ru–P
hs d^4 11110	$MnCl_5^{2-\,a}$	2.58	2.30	Mn–Cl
	$MnCl_5^{2-\,b}$	2.46	2.27	

[a] Bipyridinium counterion.
[b] Phenanthrolium counterion.

From the angular overlap results on these systems there is a rather interesting conclusion concerning the d-orbital stabilization energy of the axial MY bond in octahedral and square pyramidal environments. First we find that the total σ stabilization energy for the ls d^8 or d^9 complexes of Figure 12.6 are equal [$\Sigma(\sigma) = j(3e_\sigma - 9f_\sigma); j = 1$ for $d^9, j = 2$ for d^8]. This implies that the MY axial stabilization energies are zero in octahedral and square pyramidal complexes with this electronic configuration. This is clear to see in Figure 12.5 where the double occupation of z^2 exactly

Figure 12.6 Square planar, square pyramidal, and octahedral structures for which $\Sigma(\sigma)$ is equal for the electron configurations ls d^8 and d^9, implying long axial bonds in the five- and six-coordinate structures.

cancels the stabilization afforded the axial bond by occupation of the a_1 bonding orbital. The only stabilization associated with this system is for the b_1 orbital involved purely in M–Y_{ba} interaction. Table 12.3 gives values of the square pyramidal axial and basal MY stabilization energies for these configurations using Figure 12.6. Note that only for ls d^8 and d^9 systems is the d-orbital stabilization energy of the axial MY linkage zero. For the other two octahedral "Jahn–Teller unstable" configurations, ls d^7 and hs d^4, the octahedral axial bonds are only weakened relative to the equatorial ones.

Accordingly the distortions found in pseudo octahedral environments are much larger for d^9 systems than in ls d^7 or hs d^4 systems (Table 11.3). In general then long bonds are expected in these five- and six-coordinate structures. In practice for the Cu(II) systems it is sometimes very difficult to distinguish between four-, five-, and six-coordination.[147] The potentially six-coordinate case is often viewed in terms of the Jahn–Teller theorem. Here is a more widely applicable orbital explanation for the long bonds involved. For example the long apical distances in the square pyramidal case, while falling naturally into our scheme, are not approachable using the Jahn–Teller ideas. The z^2 and x^2-y^2 orbitals are split well apart in energy and the molecule is definitely not orbitally degenerate nor is this geometric feature traceable to a second-order effect involving mixing of these orbitals. The inherent stability of the ls d^8 square planar molecule, especially for Pt(II), is understandable on the same model. Figure 12.7 shows a mixed-valence Pt(II)/Pt(IV) species where long apical "bonds" are found around the ls d^8 Pt(II) site but equivalent bonds around the Pt(IV) ls d^6 site where the apical Pt–Y bonds have a nonzero stabilization energy. A feature complicating discussion of the d^8 case is that in the

Table 12.3   Stabilization Energies Associated with Md–Ligand σ Interaction in Square Pyramidal^a and Octahedral Molecules

Electronic Configuration	M–Y_{axial}	M–$Y_{basal\ or\ equatorial}$
hs d^4 11110	e_σ	$1.75e_\sigma$
ls d^6 22200	$2e_\sigma$	$2e_\sigma$
ls d^7 22210	e_σ	$1.75e_\sigma$
ls d^8 22220	0	$1.5e_\sigma$
d^9 22221	0	$0.75e_\sigma$
d^{10} 22222	0	0

^{a} $Y_{ap}MY_{ba} = 90°$.

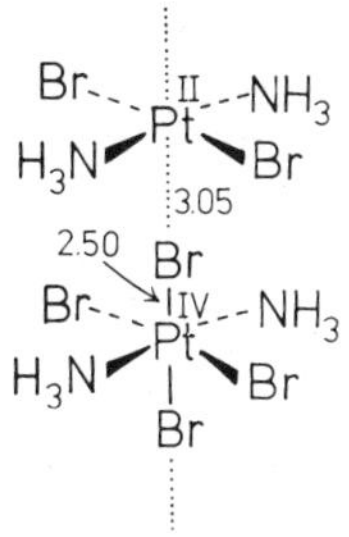

Figure 12.7 An interesting example showing long axial "bonds" around ls d^8 Pt(II) but normal axial and equatorial bonds around ls d^6 Pt(IV) in the mixed-valence species $Pt^{II}Pt^{IV}$ $(NH_3)_4Br_6$.

octahedral molecule a triplet species is possible. Here, since x^2-y^2 and z^2 orbitals are equally occupied, "apical" and "basal" bonds become equivalent and are associated with nonzero stabilization energies. One factor stabilizing this species is the exchange energy associated with the parallel spins of the triplet and the smaller coulombic repulsion energy of these two electrons when in two different orbitals. A fascinating example showing the similar stability of the high- and low-spin forms are the Lifshitz salts of Ni^{2+} with ethylenediamines. Depending upon experimental conditions of temperature, solvent, humidity, either the diamagnetic (ls d^8) form with essentially square planar coordination or paramagnetic (hs d^8) form with two short metal–solvent bonds may be made (Figure 12.8). The solution equilibria of Ni^{2+} with acetylacetonate are complex and indicate low-energy pathways connecting four, five and six coordinate molecules in accord with this idea. The greater prevalence of low-spin structures for the heavier d^8 members is explicable on the basis of larger values of Dq or e_σ and smaller values of the electron–electron repulsion parameters for these metals.

One disturbing feature of the analysis here is that the distortions in these octahedral and square pyramidal species are so large when only the d-orbital part of the stabilization has been removed. As we have emphasized, s- and p-orbital interactions with the ligands are also of importance, and especially so at the right-hand side of the periodic table. Thus $ZnCl_4^{2-}$ (d^{10}) with zero net MY d-orbital stabilization does not have abnormally long bond lengths. Longer apical or axial bonds in these systems compared to the basal or equatorial ones are expected but the very long MY distances observed are perhaps surprising.

A low-spin d^8 tbp species will contain four electrons in the e' species σ antibonding orbitals, mainly x^2-y^2 and xy. Because of the spatial extent

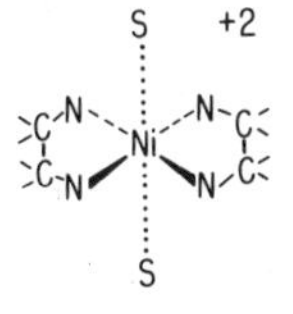

Figure 12.8 The Lifschitz salts, square planar (S absent) and diamagnetic or octahedral and paramagnetic (S = donor solvent molecule, water, or anion).

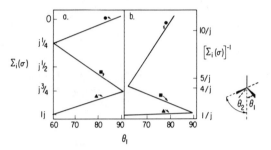

Figure 12.9 Plots of bond stabilization energy against Berry pseudorotation coordinate θ_1 ($\theta_2 = 90°$): (*a*) linear plot; (*b*) reciprocal plot. (See Figure 12.10 for definition of symbols.)

of these orbitals they are only σ antibonding between the metal and the equatorial ligands. For systems with this electronic configuration, on σ grounds longer equatorial than axial bonds are expected. We reserve comparing this result with experiment until the next chapter since these e' orbitals may also be involved in MY π interactions, and both σ and π bonding may be important here. For the d^{10} systems with x^2-y^2/xy and z^2 equally occupied, axial and equatorial bond lengths should be the same on d-orbital grounds. The relative bond lengths $r_{ax}/r_{eq} > 1$ should then be set by differential p orbital interactions. However for $CdCl_5^{3-}$ and $HgCl_5^{3-}$ with these configurations the reverse condition is found, $r_{ax}/r_{eq} < 1$. A possible explanation is suggested in the next chapter.

Many five-coordinate systems lie between the spy and tbp geometric extremes.[145] The very long apical MY bond is only associated with the spy molecule in ls d^8 and d^9 configurations. In the tbp the highest-energy (z^2) orbital (either completely or partially unoccupied) is engaged in axial and equatorial interactions. Figure 12.9 shows how the bond lengths are expected to change as the spy is distorted to the tbp, by connecting with straight lines the quadratic stabilization energies of the relevant linkages in spy and tbp molecules related by the Berry pseudo rotation coordinate.

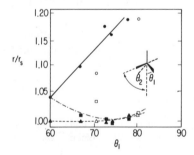

Figure 12.10 Observed bond length variations in $CuCl_5^{3-}$ units and $Ni(CN)_5^{3-}$. Berry process coordinate $\theta_2 = 90°$. Circles represent r_1; squares, average value of r_2 and r_3; triangles, average value of r_4 and r_5; solid symbols, $CuCl_5^{3-}$; open symbols, $Ni(CN)_5^{3-}$; r/r_s is the ratio of a given bond length to the value found for $r_{4,5}$ in the trigonal bipyramid or geometry closest to it. The numbering scheme is defined in Figure 6.14.

Figure 12.10 shows an experimental plot for several $CuCl_5^{3-}$ species and $Ni(CN)_5^{3-}$. The predicted trends are matched experimentally in qualitative terms. The reciprocal plot compares quite favorably with experiment. A justification for this might lie with Badger's rule which relates bond length and vibrational force constant. Force constant $= (\partial^2 V/\partial r^2)_0 = kr^{-3}$. On integration, $r \propto V^{-1}$ (V is the potential energy of the system and identified here with a bond stabilization energy).

12.2 π Bonding

The role of π orbitals of donor and acceptor type in affecting the energy of the t_{2g} set of orbitals in the octahedral complex was described in Section 9.1. Figure 12.11a shows how, on switching on the π acceptor interaction in an octahedral complex, $6c_2^2$ electrons are transferred to the π orbital of the acceptor purely by mixing of the metal d and empty ligand π orbitals. Before interaction there are six electrons in purely metal-located d orbitals. After interaction the electron density is now distributed between metal ($6c_1^2$) and ligand ($6c_2^2$) orbitals. This π back donation and the stabilization of occupied orbitals is of great importance in stabilizing many transition metal complexes. By way of contrast π donor ligands are only of importance in stabilizing octahedral systems with a small number of d electrons since for a ls d^6 system there are an equal number of MY π antibonding electrons and π bonding electrons. Perhaps the most studied π acceptor ligand is CO, where this type of interaction is usually considered to be very important. There is a large amount of crystallographic and infrared spectral data for a wide variety of these systems.

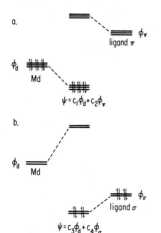

Figure 12.11 Origin of the change redistribution on complexation of (a) a π-acceptor ligand and (b) a σ-donor ligand. (Only the relevant orbitals of the octahedral complex are shown.)

A synergic bonding model is well established for terminal CO groups at least. The overlap of a carbon sp hybrid orbital and a metal d orbital with identical symmetry characteristics leads to the transfer of electron density from the carbonyl to the metal (Figure 12.11b). Also the carbonyl π and π* orbitals have the correct symmetry to overlap with metal orbitals. Three types of orbital arise from this three orbital interaction process (Figure 12.12). They are a destabilized orbital (a) largely CO π* which is antibonding between metal and carbon and between carbon and oxygen, a mainly metal-localized orbital (b) which is stabilized overall as a result of a larger interaction with π* than with π, and a rather weakly stabilized orbital (c) largely π bonding between both atom pairs. The form of these orbitals, from a recent SCF Xα calculation[106] on Cr(CO)$_6$, is shown in Figure 12.13. Orbitals b and c are usually occupied, but a is empty. Energetically the behavior of orbital b is of most importance and results in an overall stabilization of the system. Energetically this π interaction is calculated to be less important than the σ interactions in attaching the CO to the metal in this molecule but is of vital importance in determining the CO bond order, of primary interest in infrared spectral studies. The combined σ and π* effects of Figure 12.11 are usually shown in terms of the composite diagram of Figure 12.14. Although this clearly shows the charge redistribution after σ donation and π acceptance, it does not show how the stabilization occurs energetically.

For the stabilization of transition metal olefin complexes a synergetic model has also been confirmed by detailed accurate calculation after its initial suggestion by Dewar[47] and Chatt and Duncanson.[36] Figure 12.15 shows the donor and acceptor functions. The stabilization associated with the electrons in the d orbital will, from perturbation theory arguments, be inversely proportional to the d–π* energy gap. Similarly the stabilization of the electrons in the olefin π orbital will depend on the a–π energy gap. It turns out that the energy of the π* orbital varies much more from one system to another than that of the π orbital and interest-

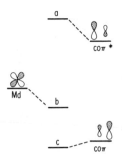

Figure 12.12 Schematic three-orbital picture showing CO coordination to a transition metal.

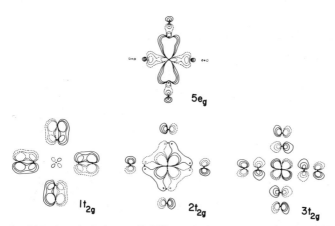

Figure 12.13 SCF $X\alpha$ calculation on $Cr(CO)_6$. 1, 2, $3t_{2g}$ correspond to the orbitals *a, b, c* of Figure 12.12; $5e_g$ is the M–C antibonding orbital of (for example) Figure 9.2. Reprinted with permission from J. B. Johnson and W. G. Klemperer, *J. Am. Chem. Soc.* **99** 7135 (1977). Copyright by the American Chemical Society.

ingly a plot of log K against $E(\pi^*)$ (from spectroscopic measurements) for the reaction

$$NiL_3 + \text{olefin} \rightleftharpoons NiL_2\text{olefin} + L \qquad (12.4)$$

shows a rough correlation[202] in agreement with these ideas (Figure 12.16).

We may derive, using the method used in the previous section, the π stabilization energies for the series (Figure 12.17) of octahedrally coordinated molecules ML_xY_{6-x} where we ignore π orbitals on Y and L is a π acceptor. In these molecules, which are particularly simple ones to look at, σ and π interactions with the *d* orbitals are separable because of the orthogonal arrangement of ligands. To low energy lie three *d* orbitals,

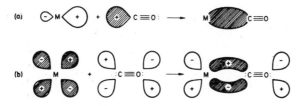

Figure 12.14 Conventional synergic bonding model for CO coordination. (In contrast to the rest of the figures in this book, the shading represents filled orbitals; the signs, the relative phases of the wavefunction.) Reprinted with permission from *Advanced Inorganic Chemistry* 3rd ed., Wiley, New York (1972).

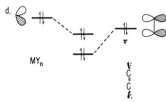

Figure 12.15 Chatt–Dewar model for ethylene co-ordination showing donor (d) and acceptor (a) orbit-als of an MY$_n$ fragment.

derived from the t_{2g} trio of the octahedron, involved purely in ML π bonding, and to higher energy lie four orbitals, derived from the e_g pair of the octahedron and involved in ML and MY σ interactions. We focus on the former and can readily describe their energies as in Figure 12.18 where n_x, n_y, and n_z are the number of π acceptor ligands with two per-pendicular π* orbitals lying along each of the x-, y- and z-axes, respec-tively. For the octahedron, for example, for the ls d^6 system, the total π stabilization energy is simply $\Sigma(\pi) = 24e_\pi - 96f_\pi$; and for the square pyramidal, $\Sigma(\pi) = 20e_\pi - 68f_\pi$. Clearly for the octahedron $\Sigma_i(\pi)$ the π stabilization associated with one ML bond is $4e_\pi - 16f_\pi$, but a little more arithmetical manipulation is needed to extract the values of $\Sigma_i(\pi)$ for the two symmetry-unrelated ML bonds of the square pyramid. We proceed as before by dividing any cross terms which arise in the evaluation of the coefficients of the f terms equally between the two sets of potentially different linkages. For example Figure 12.19 shows how the square pyr-amid axial and basal interactions are separated; xy, which is located only in the basal plane, cannot interact with π-type orbitals belonging to ligands located on the z-axis. We readily calculate

total stabilization energy of basal ligands $= 16e_\pi - 56f_\pi$ (12.5)

$$\Sigma_{ba}(\pi) = 4e_\pi - 14f_\pi$$

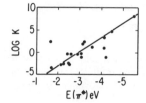

Figure 12.16 Plot of the equilibrium constant of Equation 12.4 at 25°C against the energy of the olefin π* orbital. Data from Ref. 202.

Figure 12.17 Set of octahedrally based molecules for calculation of π-bond stabilization energies showing location of ML linkages.

to be compared with

$$\Sigma_{ax}(\pi) = 4e_\pi - 12f_\pi \qquad (12.6)$$

This implies that the axial ML linkage should be stronger and shorter than the basal. In general for *all* the molecules of Figure 12.17, a simple expression describes the stabilization energy:

$$\Sigma_i(\pi) = A - (n_{cis} + 2n_{trans})B \qquad (12.7)$$

where $A = 4e_\pi - 4f_\pi$, $B = 2f_\pi$, and n_{cis} and n_{trans} are the number of ML bonds at 90 and 180°, respectively, to the one being considered. Obviously this equation shows that trans ligands lead to less overall stabilization of the system than cis ligands. This is yet another manifestation of the observation we have noted before, that two ligands in general avoid sharing the same central-atom orbital (provided the electron occupancy is such that the quadratic term in $\Sigma(\sigma, \pi)$ is angle independent). Two cis ligands share only one central-atom orbital, but two trans ligands share two (Figure 12.20). Interestingly for σ-type interactions, for these octahedrally based complexes another equation may be written:

$$\Sigma_i(\sigma) = A' - (n_{cis} + 4n_{trans})B' \qquad (12.8)$$

where $A' = 2e_\sigma - 2f_\sigma$ and $B' = \frac{1}{2}f_\sigma$. This holds for all the ls d^6 ML$_n$ species of Figure 12.17 (ignoring the Y ligands) except the T-shaped structure, where (see Section 9.1) two of the metal d orbitals mix together. In such circumstances the stabilization energy cannot so simply be apportioned between the two symmetry-unrelated linkages. On two counts then for the ls d^6 systems, ML bonds cis to one another are stronger than ML

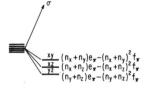

Figure 12.18 General scheme for deriving the π part of the molecular orbital diagram of the molecules of Figure 12.17. $n_{x,y,z}$ are the number of ligands along the $\pm x$, $\pm y$, $\pm z$ directions.

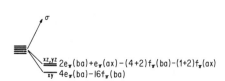

Figure 12.19 Specific example of the division of $\Sigma(\pi)$ into $\Sigma_{basal}\,(\pi)$ and $\Sigma_{apical}\,(\pi)$ for a ls d^6 square pyramidal molecule. Collecting terms in $e_\pi\,(ba)$ and $f_\pi(ba)$, the π stabilization energy is $1/4 \cdot 2[8e_\pi - 28f_\pi] = 4e_\pi - 14f_\pi$ per M–L$_{ba}$ linkage and $2[2e_\pi - 6f_\pi] = 4e_\pi - 12f_\pi$ for the M–L$_{ap}$ linkage $e_\pi\,(ax)$ and $e_\pi\,(ba)$ are of course equal, as are $f_\pi\,(ax)$ and $f_\pi\,(ba)$. We attach ax and ba labels to them to help us divide up the total stabilization energy.

bonds trans to each other in the ML$_n$ fragment. This is very neatly illustrated by the observation of the cis divacant geometry for $Cr(CO)_4$. In the ML$_x$Y$_{6-x}$ complexes the situation will be more complex since the ligand Y is a σ donor as well, but the π result still holds.

For ls d^8 systems the results will be different since now MLσ antibonding orbitals are occupied and, as seen in Section 11.4, the square planar geometry, determined by σ bonding forces, rather than the octahedral cis divacant structure is found for the four-coordinate case, for example. The results of this section hold for those systems where the quadratic energy term per ML bond is independent of coordination geometry or number of ligands and it is the quartic term which determines the relative stabilization energies. Table 12.4 shows some data for carbonyl systems; MC bonds trans to another CO are almost invariably longer than those trans to a group regarded as a poorer π acceptor, thus emphasizing the importance of π acidity in determining the structures of these molecules. In addition there are a much larger number of cis-disubstituted and fac-trisubstituted ls d^6 carbonyls than trans and mer analogs, indicating a strong energetic preference for cis rather than trans CO groups.

The inclusion of π bonding may sometimes be important in viewing the gross molecular geometries of molecules as in the previous chapter. For example ML$_5$ ls d^8 systems have a $\Sigma(\pi) = 20e_\pi - 58f_\pi$ for the tbp geometry and $20e_\pi - 68f_\pi$ for the spy structure. For the ls d^6 species however the spy structure has a larger π stabilization [$\Sigma(\pi) = 20e_\pi -$

trans

cis

Figure 12.20 d-Orbital sharing by cis and trans π-acceptor ligands.

Table 12.4 M–C Bond Lengths in $M(CO)_xX_{6-x}$ Compounds

Molecule	M–C Distance (Å)	
	Trans to CO	Cis to CO
$Cr(CO)_6$	1.909(3)	
$Cr(CO)_5PPh_3$	1.867(4)–1.894(4)	1.845(4)
$Cr(CO)_5P(OPh)_3$	1.892(5)–1.904(6)	1.861(4)
$W(CO)_5tmt^a$	1.95(2)	2.05(3)
$Cr(CO)_4dppe^b$	1.884(7)	1.831(7)
$Mo(CO)_4dppe$	2.04	1.93
$trans\text{-}Cr(CO)_4[P(OPh)_3]_2$	1.88(1)	
$fac\text{-}Cr(CO)_3(PPh_3)_3$		1.838(7)
$cis\text{-}Cr(CO)_2(PPh_3)_4$		1.817
$(CO)_5CrtriphosMn(CO)_3Br$		$1.61(5)–1.70(5)^c$
$(CO)_5CrtriphosMn(CO)_3Br$	$1.71(5)–1.88(5)^{d,e}$	$1.96^{d,e}$

^{a} tmt = Thiomorpholine-3-thione.
^{b} dppe = 1,2-Bis(diphenylphosphinoethane).
^{c} Mn–C.
^{d} Cr–C.
^{e} This structure is included to show that the M–C bond cis to CO is not always shorter than those trans to CO.

$68f_\pi$] compared to the tbp [$\Sigma(\pi) = 17e_\pi - 53.5f_\pi$] where one of the M–L π bonding orbitals is unoccupied.

The largest amount of data relevant to this problem arises via the consideration of carbonyl stretching force constants.[82] There is theoretical evidence that these are largely controlled by the extent of π back donation—the larger the magnitude of metal–COπ^* interaction, the lower the CO stretching frequency and force constant. In qualitative language the "amount" of charge available is usually regarded as dependent upon the "competition" for it by the other ligands attached to the metal. The presence of good π acceptors in addition to CO will lead to less charge available for a particular CO group, especially if the other acceptors share the same metal orbital(s). Thus in general CO groups trans to other CO groups or another good π acceptor have larger stretching force constants than CO groups trans to a vacancy (in an unsaturated system) or Y group. For binary carbonyls and carbonyl fragments $M(CO)_n$ there is a remarkably simple equation devised by Timney[201] which fits the observed force constants with incredible accuracy:

$$k_{CO} = k_d + \Sigma n_\theta b_\theta \qquad (12.9)$$

k_{CO} is the frequency-factored (Cotton–Kraihanzel)[41] "carbonyl" stretching force constant; k_d is the force constant of the monocarbonyl MCO with the same number of d electrons (Table 12.5); n_θ are the number of ligands at an angle θ to the one we are considering; and the b_θ are ligand effect constants. Thus for the square pyramidal $M(CO)_5$ molecule,

$$k_{CO}(\text{axial}) = k_d + 4b_{cis}$$

$$k_{CO}(\text{basal}) = k_d + 3b_{cis} + b_{trans}$$

(12.10)

The b parameters for CO are found experimentally to be $b_{90} = b_{cis} = 35.5 \text{ Nm}^{-1}$, $b_{180} = b_{trans} = 126.1 \text{ Nm}^{-1}$. The relative magnitudes of b_{cis} and b_{trans} are in the same order as the constants of Equation 12.7. Indeed the special case of Equation 12.9 for octahedral systems (Equation 12.11),

$$k_{CO} = k_d + (33.5n_{cis} + 126.1n_{trans})$$

(12.11)

is very similar indeed to that of Equation 12.7. There is a change of sign before the parentheses in Equations 12.7 and 12.11 since in the latter we are considering the CO linkage and in the former the MC bond. As the MC bond gets stronger so the CO linkage becomes weaker. Equation 12.9 may in fact be extended to give Equation 12.12 (Timney's master equation):

$$k_{CO} = k_d + \Sigma n_\theta(CO)\, b_\theta(CO) + \Sigma n_\theta(Y)\, b_\theta(Y) + qk'$$

(12.12)

The effect of other ligands Y on the CO stretching force constant is now included. A list of $b_\theta(CO)$ and in general $b_\theta(Y)$ parameters is given in Table 12.6. Large values are found for ligands typically regarded as good π acceptors. k' is another constant which allows for the charge q on the complex and is equal to 197 Nm^{-1}. $Mn(CO)_6^+$, $Cr(CO)_6$, and $V(CO)_6^-$ all

Table 12.5 Force Constants k_d of Metal Monocarbonyl Fragments Used in Equation 12.9 (Units Nm^{-1})

Period	k_5	k_6	k_7	k_8	k_9	k_{10}
1	1373	1387	1444	1498	1554	1610
2		1389		1506		1636?
3	1353	1381	1445	1498		1613

From Ref. 201.

Table 12.6 Timney's Ligand Effect Constants $(Nm^{-1})^a$

Ligand	$b_{90}(Y)$	$b_{180}(Y)$	$b_{tet}(Y)$	$b_{90}(Y)^b$	$b_{120}(Y)^b$
CO	33.5	126.1	37.3	25.5	51.4
PF_3	33.2	141.6	44.9	16.0	44.6
PCl_3	30.6	109.3	35.3	21	
PCl_2Ph	14	82	13.0		
$PClPh_2$	−5	55	−11.0		
PMe_3	−27.7	29.8	−38.7	−61	
PPh_3	−21	29	−31.7	−52	
$P(OMe)_3$	−15.2	66.3	−11.2	−30	
MeCN	−14	30	−21.9		
EtCN	−9	30	−22.9	−22	
BuCN	−9	31	−23.1		
$P(OPh)_3$	1.3	94	−0.3		
$As(OEt)_3$			1.9		
$AsPh_3$	−18	36	−30.9		
$SbEt_3$	−24	18	−36.1		
PCy_3			−51		
NO	42	232	30.0	45	22
CS	56	160	65		
N_2	14.0	52.0	5.7		
Cp	—	159	99		
PCl_2OEt	19.1	96.6	23		
PEt_3	−32	26		−64	
Cl	143	106	145		
Br	134	101	141		
I	112	104	125		
H	75	129			
CH_3	71	92	71		

^{a} From Ref. 201.
^{b} In trigonal bipyramid.

use the same k_d value (k_6) (Table 12.5), but these three systems have different values of the charge (q in Equation 12.12) of $+1$, 0, and -1. The sign of this charge parameter k' is readily understood. As the complex becomes negatively charged, the metal d orbitals rise in energy and move closer to the CO π^* orbitals. Accordingly their mutual interaction will be larger, and a larger "amount" of electron density will be back donated. Analogously a positive charge on the complex increases the metal

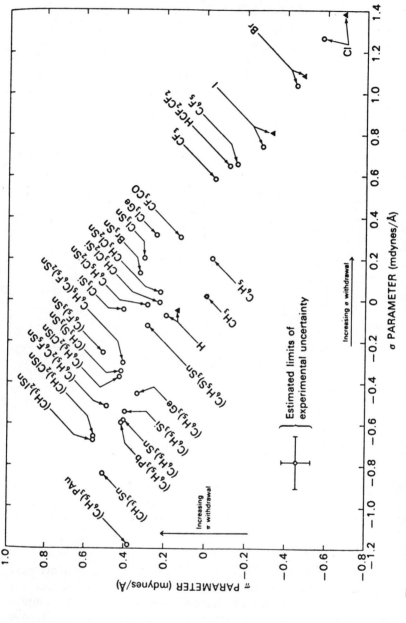

Figure 12.21 Plot of $\Delta\sigma_L$ and $\Delta\pi_L$ from Equations 12.12. Reprinted with permission from W. A. G. Graham, *Inorg. Chem.* **7** 318 (1968). Copyright by the American Chemical Society.

d–ligand π* energy gap and a smaller interaction results. The effect is often couched in phraseology which implies that with the extra electron [e.g., $V(CO)_6^-$ compared to $Cr(CO)_6$] there is more electron density available for back donation. This is not strictly true since all three of these systems are d^6 species and isoelectronic. While the functional dependence of k_{CO} on the number of cis and trans ligands can be understood using the discussion above, it is still surprising that the Timney equation holds so well and that the k_d values hold for a variety of different systems, for example, $Cr(CO)_6$, $Mn(CO)_5Br$, and $Fe(CO)_4I_2$.

A different approach has been used by Graham,[78] who has used the factor of 2, linking the π acceptor effects of cis and trans ligands, directly in the Equations 12.13:

$$\Delta k_{trans} = \Delta\sigma_L + 2\Delta\pi_L$$

$$\Delta k_{cis} = \Delta\sigma_L + \Delta\pi_L$$

(12.13)

These represent the change in CO stretching force constant for $LM(CO)_5$ species relative to a reference molecule with L = CH_3. $\Delta\sigma_L$ and $\Delta\pi_L$ represent the inductive (via a σ effect) and π contributions to the force constants and may be calculated if k_{ax} and k_{ba} are known. The result is shown in Figure 12.21. Since good σ donors (large $-ve$ $\Delta\sigma_L$) are also good π acceptors on the scheme (large $+ve$ $\Delta\pi_L$), the result was interpreted as evidence of the importance of the synergistic mechanism in the binding of π acceptors (i.e., increasing σ donation will provide more metal located electron density to be π back donated, and vice versa). However the factor of 2 in the Equations 12.13 refers to bond stabilization energies (*V*), whereas force constants are sensitive to the second derivative of *V*

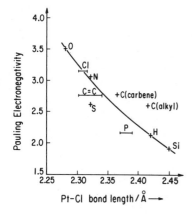

Figure 12.22 Pt^{II}–Cl bond length as a function of the electronegativity of the trans ligand.

with respect to stretching coordinate $(\partial^2 V/\partial r^2)$. Algebraically the fact that $\Delta\sigma_L/\Delta\pi_L$ of Figure 12.22 is approximately constant is equivalent to the Timney result (purely an analytical presentation of the data with no chemical bias) that b_{90}/b_{180} is approximately constant.

Although less data are available, NO and NN stretching force constants seem to be closely related to CO stretching force constants in analogous species[201] (Equations 12.14 and 12.15):

$$k_{NN}(ML_nN_2) = 1.355 k_{CO}(ML_nCO) - 210 \quad (\text{in Nm}^{-1}) \quad (12.14)$$

$$k_{NO}(M^{-1}L_nNO) = 0.8549 k_{CO}(ML_nCO) - 54 \quad (\text{in Nm}^{-1}) \quad (12.15)$$

A less general Equation (12.16) holds for the CS ligand:

$$\nu_{CS}(M(CO)_nCS) = 0.4262 k_{CO}(M(CO)_{n+1}) + 604 \quad (\text{in cm}^{-1}) \quad (12.16)$$

12.3 The Trans Influence

A pervading theme of the previous two sections has been the influence on the stabilization energy of a given metal-ligand linkage of other ligands coordinated to the metal atom. A specific and well studied facet of this mutual influence of ligands is the trans influence, i.e.[88,137,171,208] the effect on MY bond properties as a function of the nature of the ligand trans to Y in the complex. We distinguish immediately between the trans-influence, a measure of a (static) bond property (length, vibrational force constant or frequency, magnetic resonance properties of Y etc.) and the (kinetic) trans-effect (which we will not discuss), the dependence of the rate of substitution of the ligand Y by an incoming nucleophile on the nature of the trans ligand.[125] Many of the examples studied have been square planar ls d^8 Pt(II) complexes. Here the four ligands are attached to the metal by interaction with the x^2-y^2 orbital augmented by interaction with the higher energy metal 6s and 6p orbitals. In a square planar $MYTC_2$ complex (T and C represent ligands trans and cis to Y, respectively) the MY bond length is much more sensitive than the MC bond lengths to variations in the nature of the trans ligand T (Tables 12.7 and 12.8) suggesting that the trans influence is communicated predominantly via the unidirectional 6p orbital rather than by the x^2-y^2 or 6s orbital which may be involved equally with all four ligands.

Existing approaches to the trans influence include "competition" for the available metal p-orbital character by the ligands Y and T. Here we briefly examine this aspect of this structural observation using the AOM

Table 12.7 The Trans Influence of Ligands on Platinum(II)–Chlorine Bond Lengths

Complex	Trans Atom (or Ligand)	Pt–Cl (Å)

Pr₃P   NCS   Cl
  \ /   \ /
   Pt   Pt
  / \   / \
Cl   SCN   PPr₃
	—NCS	2.277 (4)

K[Pt(acac)₂Cl]	O	2.28 (1)
trans-[(PEt₃)₂PtCl]	Cl	2.30 (1)
cis-[(p-C₆H₄S)₂PtCl₂]	S (of RS⁻)	2.30
trans-[(PEt₃)₂Pt(CO)Cl]	CO	2.30
(PPh₄)[PtCl₃(HOCH₂CH=CHCH₂OH)]	C=C	2.301 (8)

Pr₃P   SCN   Cl
  \ /   \ /
   Pt   Pt
  / \   / \
Cl   NCS   PPr₃
	—SCN	2.304 (4)

cis-[(PEt₂PH)Pt(CNEt)Cl₂]	RNC	2.314 (10)
K₂[PtCl₄]	Cl	2.316
[Pt(L-methionineH)Cl₂)]	S (of R₂S)	2.32
K[Pt(NH₃)Cl₃]H₂O	N	2.321 (7)
K[Pt(C₂H₄)Cl₃]H₂O	C=C	2.327 (7)
[Pt(H₃NCH₂CH=CHCH₂NH₃)Cl₃]Cl	C=C	2.342 (2)
cis-[PEt₃)Pt{C(OEt)NHPh}Cl₂]	C (of carbene)	2.365 (5)
cis-[(PMe₃)₂PtCl₂]	P	2.37 (1)
cis-[(PEt₂Ph)Pt(CNEt)Cl₂]	P	2.390 (8)
trans-[(PMe₂Ph)₂Pt(CH₂SiMe₃)Cl]	C (of alkyl)	2.415 (5)
trans-[(PPH₂Et)₂PtHCl]	H	2.42 (1)
cis-PtCl₂ { CH₂ṄH₂CH(Me)Ph / CH \ CH₂CH₂CH=CH₂ }	C (of alkyl)	2.430
trans-[(PMe₂Ph)₂Pt(SiMePh₂)Cl]	Si	2.45 (1)

From Ref. 88.

Table 12.8   The Cis Influence of Ligands on Platinum(II)–Chlorine Bond Lengths

Molecule	Cis Atom (or Ligand)	Pt–Cl (Å)
trans-[(PEt$_3$)$_2$PtCl$_2$]	P	2.29
[Pt(*cis*-MeCH=CHCH$_2$NH$_3$)Cl$_3$]	C=C	2.297 (6)
trans-(PEt$_3$)Pt $\left\{ \begin{array}{c} Ph \\ N \\ / \quad \diagdown CH_2 \\ C \quad\quad | \\ \diagdown \quad \diagup CH_2 \\ N \\ Ph \end{array} \right\}$ Cl$_2$	P and C (of carbene)	2.30
K[Pt(C$_2$H$_4$)Cl$_3$]H$_2$O	C=C	2.305
K[Pt(NH$_3$)Cl$_3$]H$_2$O	N	2.315
K$_2$[PtCl$_4$]	Cl	2.316
trans-[Pt(NH$_3$)$_2$Cl$_2$]	N	2.32 (1)

From Ref. 88.

but point out that this may not be the only factor determining the relative bond lengths in these molecules. For two ligands Y and Y′ sited trans to one another use of the angular overlap ideas leads to a very simple form of the bond stabilization energy via this orbital-sharing process. The difference in MY stabilization energy on going from MY$_2$ to MYY′ is simply[31] given by Equation 12.17, where Δe_σ is the difference in the e_σ values of Y and Y′:

$$\Delta \Sigma_{MY}(\sigma) \propto e_\sigma \cdot \Delta e_\sigma \qquad (12.17)$$

So if the Y′ ligand has a larger e_σ value than the Y ligand, the MY bond is predicted to be weakened compared to MY$_2$. If Y′ has a smaller e_σ value, then the MY bond is predicted to be strengthened. Since a smaller e_σ value is associated with increasing ligand electronegativity, the MY bond length should increase with decreasing electronegativity of the trans ligand T = Y′. Figure 12.22 shows a plot of the experimental data; the trend is quite a striking one. Another facet of the trans influence is that the best trans influencing ligands are often the ones most easily influenced. This also follows from Equation 12.17. The effect is not only proportional to the difference e_σ values between the ligands Y and T (Δe_σ) but also the absolute value of e_σ for M–Y interaction itself.

13 More Detailed Form of the Molecular Orbitals

Up until now we have mainly used a d orbital-only model to describe transition metal chemistry, with the effects of the higher-energy s and p orbitals added on where needed. Most of the arguments have been symmetry or orbital overlap based, and we have found the AOM a very useful way of making these semiquantitative. However the AOM does not treat the problem of actual hybridization of these orbitals very cleanly. In order to look in more detail at the molecular orbital basis of transition metal chemistry, we need to use more qualitative, nonparametrized ideas, supported by the results of quantitative calculations.

13.1 s, p, d Mixing

The hybridization, or mixing, of these orbitals drops naturally out of the molecular orbital model if the symmetry of the system is such that one of the metal d orbitals transforms as the same symmetry species as either the metal s or one of the metal p orbitals, or both. In the octahedron the orbital symmetries are such that $d–p$ or $d–s$ mixing may not occur. Mixing may occur however in the lower-symmetry square pyramidal molecule (Figure 13.1). Here the result of mixing the higher energy s and p_z orbitals into "z^2" of the d orbital-only model leads to the synthesis of a hybrid orbital which points to the vacant, sixth coordination site of the octahedron. The relative phases of mixing the three orbitals together are readily appreciated from simple perturbation theory arguments. The energy of the deepest-lying member of the set (s, p, d) will be pushed to lower energy by such interactions, that is the z^2 orbital will be "repelled"

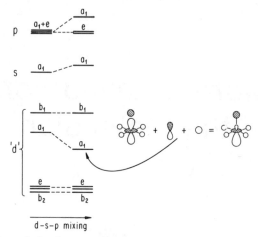

Figure 13.1 Effect of allowing central-atom d, p, and s orbitals of a_1 symmetry to mix together in a square pyramidal MY_5 molecule. The relative phases of s and p mixing into z^2 are also shown.

by the higher-lying s or p orbitals. This requires that the higher-energy orbitals mix into z^2 in a bonding fashion so as to minimize hybrid–ligand antibonding. Second-order perturbation theory shows in more detail how the phase of this mixing process is determined. The first-order correction to the wavefunction (Equation 1.10) will only describe how two orbitals mix together, the second-order correction will allow the description of a three-orbital mixing process. With reference to Figure 13.2 the coefficient describing the amount of p_z mixed into z^2 is proportional to

$$\frac{S(z^2, \sigma)S(p_z, \sigma)}{(E_{z^2} - E_\sigma)(E_{z^2} - E_{p_z})} \tag{13.1}$$

where the overlap integrals S are between metal d or p orbitals and the ligand σ orbitals. These are chosen arbitrarily to be positive (although it makes no difference to the argument). Since the energy levels lie in the

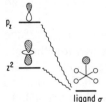

Figure 13.2 Pertubation analysis of the phase of mixing of Figure 13.1 using the metal p orbital as an illustration.

Figure 13.3 *d,p* Mixing in square pyramidal MY$_5$ molecules as θ, the Y$_{ap}$MY$_{ba}$ angle, changes from 90°.

order $E_\sigma < E_{z^2} < E_{pz}$, the denominator of Equation 13.1 is negative and p_z mixes into z^2 with a minus sign. The extent of hybridization is clearly proportional to the overlap integral and inversely proportional to the energy separation from the ligand σ orbitals for each of the two relevant metal orbitals. The effect in the square pyramid is to depress the energy of the a_1 z^2 orbital. (An analogous effect is seen in a C_{3v} pyramid (MY$_4$) when p_z–z^2 mixing is switched on as a result of removal of an axial ligand from a trigonal bipyramidal MY$_5$; z^2–s mixing is allowed in the tbp.) At the geometry where the Y$_{ba}$MY$_{ax}$ angle = 90°, there is no mixing between the *d* orbitals of *e* symmetry (*xz, yz*) and $p_{x,y}$ (also of symmetry species *e*). The *p* orbitals may only interact with ligand σ orbitals and the *d* orbitals of this symmetry type only with ligand π orbitals, so that there is no mechanism for mixing. As this YMY angle changes however, *d–p* mixing may occur, and the directions of the hybrid orbitals that result are shown in Figure 13.3.

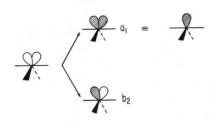

Figure 13.4 Localized and symmetry orbitals from an octahedral cis divacant fragment.

Figure 13.5 Localized and symmetry orbitals from an octahedral fac trivacant fragment.

Generally removal of ligands from the octahedral structure leads to the generation of orbitals which on a localized basis point toward the vacant sites (Figures 13.4 and 13.5). On a delocalized basis they may be viewed as symmetry-adapted linear combinations.

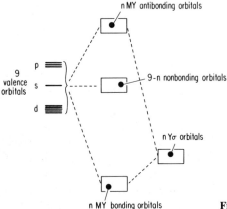

Figure 13.6   Origin of the l8-electron rule.

13.2   The Number of Valence Orbitals and the 18-Electron Rule[142]

Each metal atom has nine valence orbitals (five d, three p, and one s). In an MY_n complex, where each ligand carries a single σ orbital, each of the collection of n symmetry orbitals will find in most geometries one among these metal orbitals with which to interact (Figure 13.6). Thus n of the nine are destabilized by σ interaction, the n ligand σ orbitals are doubly occupied and pushed to lower energy leaving $9 - n$ metal orbitals, either to hold metal d electrons or to act as donor orbitals which may be stabilized by higher-energy orbitals on the ligands. With a total of 18 electrons associated with the metal coordination, all the n ligand-located σ-bonding orbitals and all the lower-lying valence orbitals are filled. A large energy gap between these filled valence orbitals and the metal orbitals involved in metal–ligand σ antibonding interactions bestows structural and kinetic stability on the molecule. Clearly the 18-electron (or effective atomic number) rule which very usefully controls the stoichiometry of many transition metal complexes will hold in those cases where the HOMO–LUMO gap is large. This will be the case for ligands that are good σ donors (and therefore destabilize the LUMO) and good π acceptors (and therefore depress the energy of the HOMO). Thus complexes containing high Δ ligands (CO, phosphines, ethylene, etc.) at one end of the spectrochemical series will adhere to the rule that the total electron count per metal atom will be 18 electrons in stable complexes, more than low Δ ligands such as H_2O and halogen. So the majority of stable organometallic compounds containing one or more high Δ ligands invariably have an 18-electron count at the metal atom (e.g., $Cr(CO)_6$, $FeCp_2$, $Ni(PF_3)_4$, $CpMn(CO)_3$); but species such as $FeCl_4^{2-}$ (14 electrons) and

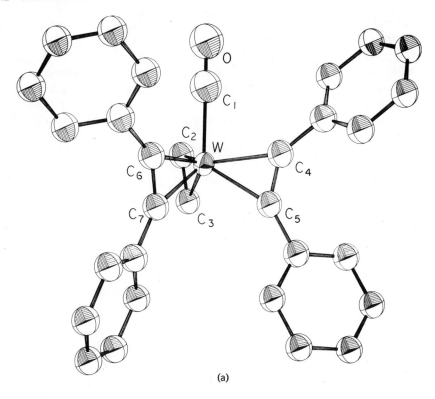

(a)

Figure 13.7 (a) The W(CO) (C₂Ph₂)₃ molecule (reprinted with permission from L. M. Laine, R. E. Moriaty, and R. Bau, *J. Am. Chem. Soc.* **94** 1402 (1972). Copyright by the American Chemical Society). (b) The ligand a_2 π orbital combination which finds no central atom counterpart in this apparently 20-electron molecule.

$Zn(H_2O)_6^{2+}$ (22 electrons) also exist. Several organometallic species containing metal atoms from the left hand side of the periodic table also contain fewer than 18 electrons.

One exception to the rule is the fact that many complexes of the d^8 metals are low spin and square planar but have only 16 electrons (e.g., $Ni(CN)_4^{2-}$). The special stability of this geometry may be traced to two reasons: (a) The large energy gap between z^2 and x^2-y^2 orbitals (Figure 9.7); this is accentuated by z^2-s mixing which depresses z^2 even further (Section 9.3). (b) There is one central-atom orbital, namely p_z, which cannot interact with the ligand σ orbitals in the square planar (or trigonal planar) geometry; this orbital is however simply at too high an energy to be filled. The fact that an extra one or two ligands may be added to produce 18-electron five- or six-coordinate compounds (or alternatively the reverse—loss of ligands from these five- and six-coordinate species) makes this particular geometry an interesting one from the point of view of catalytic processes where flexibility in coordination number is an important requirement.

Another exception to the 18-electron rule is the complex $WCO(C_2Ph_2)_3$ with 20 electrons (Figure 13.7). It is a rare example of the coordination geometry being such that one pair of electrons lies in an orbital which finds no counterpart on the metal (a_2). Cp_2Ni contains 20 electrons also. Here the two extra electrons occupy an e_{2u} orbital located on the ligands, a symmetry species not found among the central-atom orbitals. When

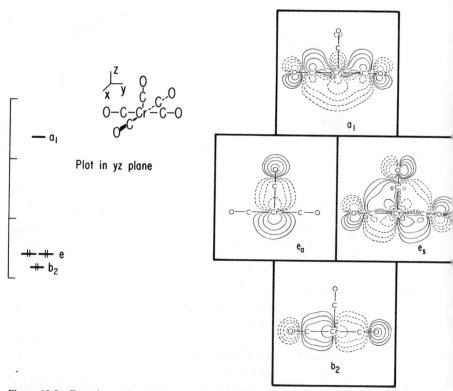

Figure 13.8 Energies and spatial orientation of the $9 - n$ orbitals of some transition metal fragments of interest. Orbitals which are antisymmetric with respect to a plane are plotted on a level of 0.5 Å above and parallel to this plane. The top views of the $Fe(CO)_3$ orbitals were plotted 1.5 Å away from the metal atom in the xy plane. The different phases of the wavefunction are represented by full and dashed lines. From Ref. 4.

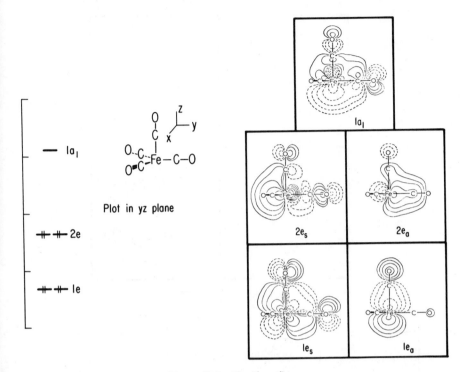

Figure 13.8 (Continued)

viewed in these terms they are not exceptions to the general theory after all. There are of course several other non-18-electron species that owe their stability to the presence of sterically bulky groups which prevent coordinative saturation. In some other species (e.g., $Cp_2Mo(NO)R$ and $Cp_2W(CO)_2$) with an apparently 20-electron count, the Cp ring has partially "slipped" off the metal to relieve this problem.

One useful item we shall need later is how many of these $9 - n$ valence orbitals lie to low energy and how many lie to high energy.[54] For geometries that lie close to that of the octahedron, the fragment retains the general orbital pattern of that geometry, and three orbitals are found to low energy (derived from the t_{2g} trio) and then $6 - n$ are found to higher energy. For rather different geometries the orbital energies will change (see for example the behavior of the spy MY_5 orbitals on bending, Figure 9.8). Figure 13.8 shows the approximate orbital energies and density plots of these $9 - n$ "frontier" orbitals for various important fragments which are useful in understanding the properties of transition metal complexes.[4]

An interesting observation is the close resemblance between the orbitals of some transition metal fragments and those of main group systems.

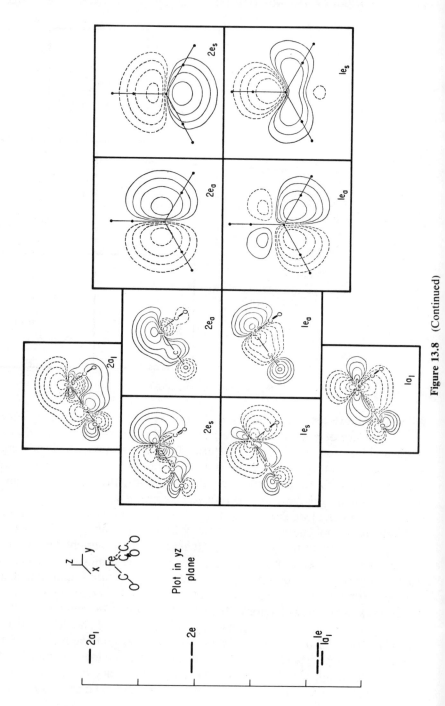

Figure 13.8 (Continued)

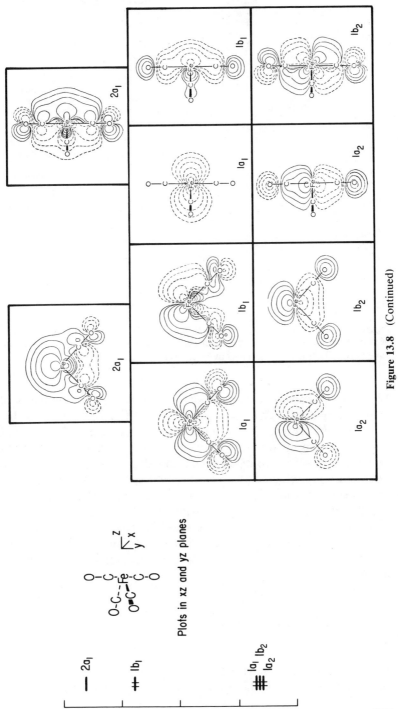

Figure 13.8 (Continued)

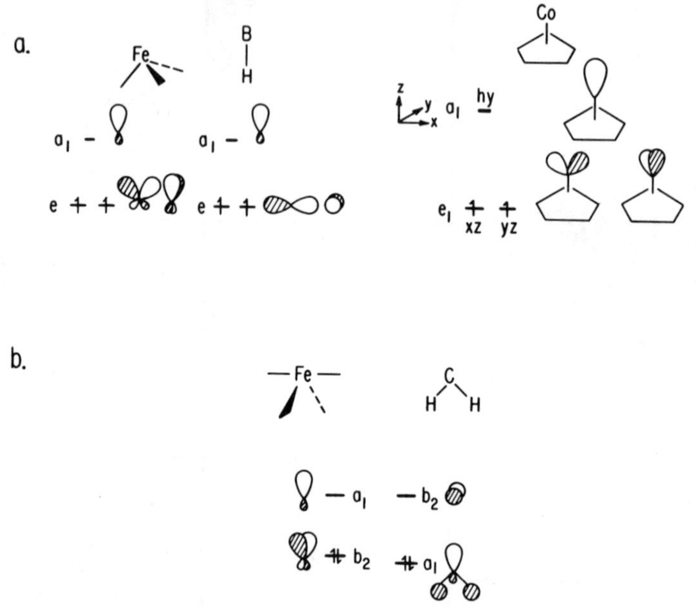

Figure 13.9 (*a*) Isolobality of Fe(CO)$_3$, BH, and CoCp. (*b*) Similarity between the orbitals of Fe(CO)$_4$ and CH$_2$. (Some of the lower-energy *d* orbitals of Fe(CO)$_3$, Fe(CO)$_4$, and CoCp have been left off these diagrams.) hy = hybrid.

Figure 13.9 shows that both in spatial description and numbers of electrons pyramidal Co(CO)$_3$ is rather like linear CH and Fe(CO)$_3$ is like BH. Thus these units in addition to being isoelectronic are also *isolobal*. This resemblance forms the basis[141] for the rationalization of the structures of polyhedral systems containing CH and BH units (carboranes), M(CO)$_x$ units (carbonyl cluster compounds), and species containing both structural units (metallocarboranes), which are examined in the next chapter. Another important unit isolobal with M(CO)$_3$ is πCpM$^-$, where in some ways the π-cyclopentadiene ring may be envisaged as occupying three (fac) coordination sites of the octahedron (Figure 13.9). Another useful pair of species not strictly isolobal but very similar in orbital structure are Fe(CO)$_4$ and CH$_2$.

13.3   The Fragment Formalism

We are familiar with the construction of molecular orbital diagrams from a central atom plus a collection of ligands. An alternative way to proceed is to combine the orbital patterns of two molecular fragments to

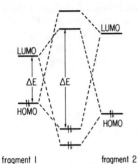

Figure 13.10 General strategy for the stabilization of two fragments on interaction, resulting in an increase in ΔE, the HOMO–LUMO gap, and a stabilization of occupied orbitals.

produce the same end product. The difference lies in the fresh light that is often shed on the nature of the interaction. Figure 13.10 shows the general requirements[62] for an effective stabilization on formation of the new molecule from the two fragments. One or more filled, low-lying orbitals of each fragment need to find an empty, close, higher-lying orbital on the other to (a) produce overall a set of occupied MOs at lower energy and (b) an increased HOMO–LUMO gap (ΔE of figure 13.10). Initially as a very simple example we consider an octahedral cis divacant ls d^6 M(CO)$_4$ fragment plus two CO ligands (Figure 13.11). It is clear that conditions are just right for the formation of octahedral Cr(CO)$_6$. Formation of Fe(CO)$_6$ from ls d^8 Fe(CO)$_4$ is an energetically unfavorable process

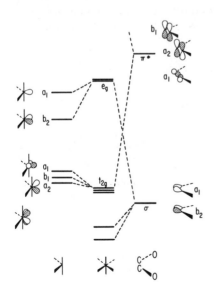

Figure 13.11 "Synthesis" of an M(CO)$_6$ molecule from M(CO)$_4$ + 2 CO. (One CO π^* combination, of b$_2$ symmetry, is not shown.)

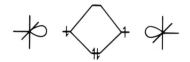

Figure 13.12 Schematic molecular diagram showing formation of $Mn_2(CO)_{10}$ from two $Mn(CO)_5$ molecules (or $Co_2(CN)_{10}^{6-}$ from two $Co(CN)_5^{3-}$ species) by overlap of two "z^2" orbitals which point out of the bottom of the spy unit (Figure 13.1).

since two electrons in the b_2 orbital will be pushed to higher energy. In all of the molecular dissections using this approach, we use for the geometry of each fragment the one which it adopts in the complex, rather than the one found in the free molecule if it exists. Thus free $Fe(CO)_4$ used in this example does not have the cis divacant structure nor does it exist as a singlet species.

The unstable species $Mn(CO)_5$ and $Co(CO)_4$ have been made in low-temperature matrices, but manganese and cobalt carbonyls are usually found as the binuclear species $Mn_2(CO)_{10}$ and $Co_2(CO)_8$. Although the behavior of the mononuclear species fits with the predictions of the 18-electron rule (in addition to dimerization the species will abstract halogens to form $M(CO)_xX$ species), an alternative way of viewing the dimerization process is to look at the form of the orbital diagram of the fragment. The HOMO is ideally spatially located (Figure 13.1) for combination with another $Mn(CO)_5$ molecule. The highest-lying electron in both fragments is stabilized on such an interaction (Figure 13.12).

The tendency of this sixth-site orbital to interact with other ligands is beautifully illustrated[23,26] by the visible spectra of $M(CO)_5$ species (M = Cr, Mo, or W) trapped in low-temperature matrices. When isolated in ostensibly "inert" matrices such as Ar, Kr, CH_4, or SF_6, the visible band

Table 13.1   Dependence of Visible Absorption Band of $Cr(CO)_5$ (ls d^6) on the Sixth-Site Occupant

	Wave length (nm)	Energy (cm^{-1})
Ne	624	16,000
SF_6	560	17,900
CF_4	547	18,300
Ar	533	18,800
Kr	518	19,300
Xe	492	20,300
CH_4	489	20,400

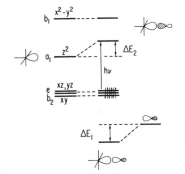

Figure 13.13 Sensitivity of the $(xz,yz) \to z^2$ ($^1A_1 \to {}^1E$) electronic transition in square pyramidal Group VI $M(CO)_5$ molecules to the nature of the species close to the sixth site.

of this species moves considerably as a function of matrix material (Table 13.1). One mechanism for these shifts is shown in Figure 13.13. The size of the interaction (ΔE) and hence the energy of the absorption band increases as the ionization potential of the sixth site occupant decreases in accordance with perturbation theory ideas. This is not the only contribution to the changes found for $h\nu_{vis}$. The e–z^2 energy gap also changes with bond angle, a process which was shown in Figure 9.8. The result does show that the "driving force" of the 18-electron rule is quite a potent one in these species. A similar effect is seen[113] in 1s d^7 Co(CN)$_5^{3-}$ and Co(CNR)$_5^{2+}$ molecules. The truly five-coordinate molecules are yellow but the crystal structure of the green form of Co(CNC$_6$H$_5$)$_5^{2+}$ (and perhaps Co(CN)$_5^{3-}$) shows a close species (perchlorate) in the sixth site. Here the perturbation is probably greater than in the carbonyl example. The Co–C apical bond length (and apical–M–basal bond angle) are 1.95 Å (95°) and 1.88 Å (101.8°) in "complexed" and "free" ions, respectively.

Mn$_2$(CO)$_{10}$ is yellow in color, and the lowest-lying electronic absorption is associated with a transition from σ to σ^* leading to fission of the Mn–Mn bond and generation of the Mn(CO)$_5$ radical. Obviously the excited Mn$_2$(CO)$_{10}$ molecule has an 18-electron count at each metal atom, but it is important that the electrons occupy the lowest-energy orbitals for the species to be stable. Analogously Cr(CO)$_6$ and Fe(CO)$_5$ decompose readily by CO loss on electronic excitation within the d manifold of orbitals. The 16-electron square pyramidal M(CO)$_5$ molecules (M = Cr, Mo, or W) however rearrange on visible photolysis via an intramolecular process rather than lose a CO group (see Section 11.5). In the excited state the promoted electron occupies the same z^2 orbital that acts as the HOMO in Mn(CO)$_5$. Higher-energy excitation leads to CO loss. Dissociative instability is then very much associated with the occupation of high-energy orbitals of the molecule and, in the cases we have described, with occupation of orbitals higher in energy than the $9 - n$ metal-located orbitals.

a   b   c

Figure 13.14 Three possible orientations of the olefin in $Fe(CO)_4$ olefin complexes.

13.4 The Trigonal Bipyramidal MY_5 Molecule[180]

There are several examples we could choose to illustrate further the use of the fragment formalism. A particularly instructive one which will eventually allow completion of our discussion of the MY bond strengths of the tbp is to look at an octahedrally based $Fe(CO)_4$ unit as in Figure 13.11 and see how it may be stabilized by the addition of a fifth ligand. There are many $Fe(CO)_4$ olefin complexes available and it is interesting that the structural preference of the olefin is in an equatorial site with the plane of the double bond in the equatorial plane of the molecule (Figure 13.14a). Alternative orientations in an equatorial site but parallel to the threefold axis of the tbp, or in an axial site, are not energetically favored (Figures 13.14b and c). Axial olefins are only found if the equatorial sites are blocked or already occupied by olefins. This result drops out naturally

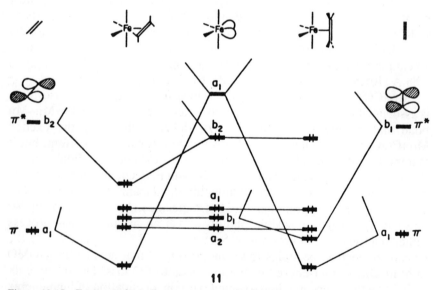

Figure 13.15 Fragment formalism approach to distinguish between the virtues of *eq* and *eq* for the olefin geometry in $Fe(CO)_4$ olefin. Reprinted with permission from R. Hoffmann, T. A. Albright, and D. L. Thorn, *Pure Appl. Chem.* **50** 3 (1978).

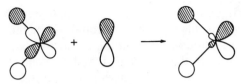

Figure 13.16 p–d Hybridization in the trigonal bipyramidal MY_5 molecule leading to superior eq π-type interaction.

from the fragment approach as shown in Figure 13.15. The $Fe(CO)_4$ orbitals are oriented in just the right way and contain just the right number of electrons to stabilize an ethylene molecule. In both orientations the a_1 (π) orbital is stabilized by approximately the same amount, but the equatorial π^* interaction b_2 is favored over the b_1 by both a larger overlap and a smaller energy separation in a perturbation ($S^2/\Delta E$) treatment of the interaction. The larger overlap is intimately connected with the hybridization of the x^2–y^2, xy orbitals with p_x, p_y as in Figure 13.16. This means that the $eq_\perp$ π-type interaction (Figure 13.14) will be larger than any axial or $eq_\parallel$ interaction. For low-spin d^8 systems π donors will avoid the equatorial position (since this will result in a destabilization of the occupied pair of xy, x^2–y^2 orbitals which contain two electrons), but π acceptors

Table 13.2 MY Bond Lengths in Some MY_5 Structures

Molecule	Configuration[b]	M–Y (Å)	
		Axial	Equatorial
$Fe(N_3)_5^{2-}$	d^5 (hs)	2.04	2.00
$Co(C_6H_7NO)_5^{2+}$	d^7 (hs)	2.10	1.98
$Ni(CN)_5^{3-a}$	d^8	1.84	1.99, 1.91
NiP_5^{2+}	d^8	2.14	2.19
$Fe(CO)_5$	d^8	1.81	1.83
$Co(CNCH_3)_5^+$	d^8	1.84	1.88
$Pt(SnCl_3)_5^{3-}$	d^8	2.54	2.54
$Pt(GeCl_3)_5^{3-a}$	d^8	2.40	2.43
$Mn(CO)_5^-$	d^8	1.82	1.80
$CuCl_5^{3-}$	d^9	2.30	2.39
$CuBr_5^{3-}$	d^9	2.45	2.52
$CdCl_5^{3-}$	d^{10}	2.53	2.56
$HgCl_5^{-3}$	d^{10}	2.52	2.64
AsF_5	d^{10}	1.71	1.66

[a] C_{2v} structure intermediate between D_{3h} and C_{4v}.
[b] The ligand is 2,8,9-trioxa-1-phosphaadamantane.

will seek it out. In addition if the acceptor contains a single π^* orbital, as in ethylene, then the plane of the π^* orbital will be the equatorial plane of the molecule. In addition this p–d mixing also means that the σ antibonding interaction with these orbitals is reduced and the importance of σ effects in determining site preferences not as strong as may have been envisaged.

Thus in a 1s d^8 ML_5 species (the L are π acceptors) the axial bonds are predicted to be weaker than the equatorial ones on π-bonding grounds. The reverse order is predicted on σ grounds (Section 12.1) and for π-donor ligands. Similar arguments hold for the d^9 series. Some results for 1s d^8 and d^9 molecules are shown in Table 13.2 and in general $r_{ax} > r_{eq}$ for the π-acceptor systems and $r_{ax} < r_{eq}$ for π donors, although often the bond length differences are small.

For d^{10} systems with all the MY σ and π antibonding orbitals filled, the relative bond lengths should adopt the ratio found for systems where s, p orbitals control the geometry (i.e., $r_{ax} > r_{eq}$). However for the d^{10} systems for Table 13.2 the reverse is true, the difference being larger for the heavier mercury system. Incipient two-coordination is a feature of structural mercury chemistry (and also of several other heavy elements) that has been rationalized over the years by z^2–s mixing,[52] particularly favored on energy gap grounds at the bottom of the periodic table. Unfortunately the tbp $ZnCl_5^{3-}$ molecule to complete the series of Table 13.2 is as yet not available.

13.5   Fe(CO)₃ and Dienes

Another example of the use of the fragment formalism shows[54] why $Fe(CO)_3$ prefers to form a variety of stable complexes with conjugated rather than with unconjugated dienes. Figure 13.17 shows the orbital scheme for $Fe(CO)_3$ alongside the diagram for three different diene structures. (Note that as for the $Fe(CO)_4$ example, the geometry of singlet $Fe(CO)_3$ is not pyramidal and that the lowest electronic state of free $Fe(CO)_3$ is probably a triplet. Singlet $Fe(CO)_3$ as we have drawn it here is Jahn–Teller unstable at this geometry.) The conjugated diene is in a position to provide a good donor and good acceptor orbital (e.g., butadiene), each of the correct symmetry to interact with one component of the higher-energy e $Fe(CO)_3$ orbital, or as in the case of cyclobutadiene to provide an orbital where there will be net transfer of two electrons to this orbital from $Fe(CO)_3$. The complexed cyclobutadiene will have properties approaching those of an aromatic 6π delocalized system as found experimentally.

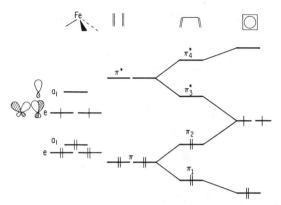

Figure 13.17   Interaction scheme for Fe(CO)₃ with dienes. Adapted from Ref. 53.

Thus the role of double bond conjugation in complex stabilization is to lower the energy of the olefin LUMO and increase the energy of the HOMO so that Fe(CO)₃–ligand interaction is larger.

13.6   Linear and Bent Nitrosyl Groups

In this section we view in molecular orbital terms the interesting structural observation that coordinated NO groups sometimes give rise to linear MNO units and sometimes to bent ones.[56,98] Some examples are shown in Figure 13.18, and the structural results are very dependent on electronic configuration. We will use the fragment formalism to generate a molecular orbital diagram for the spy system and then use overlap arguments to derive a Walsh diagram for the MNO bending coordinate. It has become common practice to regard the linear nitrosyls as containing NO^+ and bent nitrosyls as containing NO^-. (The infrared $\nu(NO)$ frequencies of bent

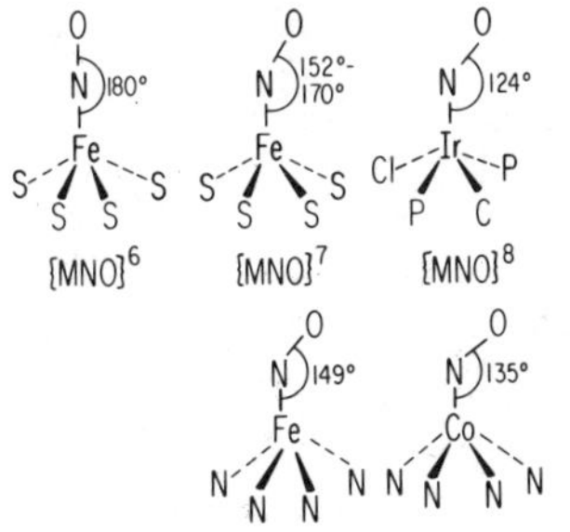

Figure 13.18   Some examples of linear and bent nitrosyl groups: $\{MNO\}^6$ Fe(NO) (S₂C₂(CNR₂)₂; $\{MNO\}^7$ Fe(NO) (S₂CNR₂)₂, Fe(NO)(tetraphenylporphyrin); $\{MNO\}^8$ Ir (PPh₃)₂(CO)(NO)Cl⁺, Co (NO)(tetraphenylporphyrin).

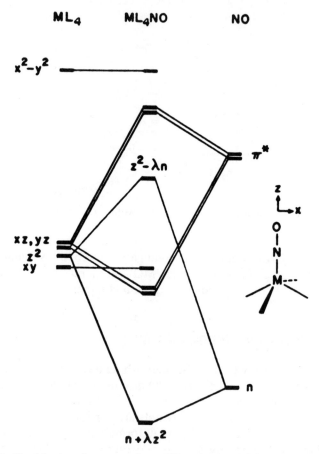

Figure 13.19 Combination of a square planar MY_4 unit with a linear apical nitrosyl ligand. Reprinted with permission from R. Hoffmann, M. M. L. Chen, M. Elian, A. R. Rossi, and D. M. P. Mingos, *Inorg. Chem.* **13** 266 (1974). Copyright by the American Chemical Society.

nitrosyls are usually much lower than those for linear species.) Such a viewpoint will not be considered here—whether a MNO unit is linear or bent will be determined by the stabilization or otherwise of occupied molecular orbitals on bending.

The important orbitals of NO to consider are the σ and π^* ones, just as in the CO case. One difference between the NO and CO units is that in the former the π^* levels are probably much closer in energy to the metal d orbitals than for CO making it a better π acceptor. Another difference is that the π^* level is singly occupied in free NO but empty in free CO. As a useful way to describe the electron configuration of the

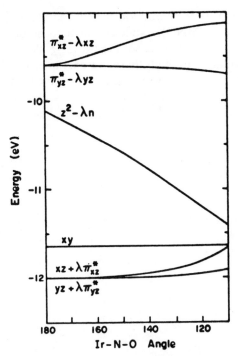

Figure 13.20   Effect on the square pyramidal (C_{4v}) MY_4NO energy levels on bending the coordinated nitrosyl. Reprinted with permission from R. Hoffmann, M. M. L. Chen, M. Elian, A. R. Rossi, and D. M. P. Mingos, *Inorg. Chem.* **13** 266 (1974). Copyright by the American Chemical Society.

complex we use the abbreviation $\{MNO\}^n$, where n is the number of electrons associated with the metal d orbitals and the π^* orbitals of the nitrosyl. Figure 13.19 shows an energy diagram for interaction of an apical, linear NO group with the fragment orbitals of a square plane. Two strong interactions result—the stabilization of the NO σ orbital by z^2 and the stabilization of the metal xz, yz pair by π-type interaction with the NO π^* orbital. The actual level ordering in this diagram is crucial to the success of the scheme. For an $\{MNO\}^8$ species the orbitals of this diagram are filled through the $z^2-\lambda n$ hybrid orbital. The highest occupied orbital is clearly destabilized relative to the planar molecule and the system energetically would profit from a distortion that lowered its energy. On bending the MNO group at the nitrogen atom, this is just what happens (Figure 13.20). The $z^2-\lambda n$ orbital is dramatically stabilized since on bending z^2 may now overlap in phase with one component of the close π^* orbital. Concurrently however one of the π-bonding orbitals encounters a destabilizing effect as π interaction is reduced. Thus $\{MNO\}^6$ species ought to be linear and $\{MNO\}^{7,8}$ species ought to be bent, with a larger distortion for the eight-electron case. Figure 13.18 shows that this can be the case experimentally although there are examples of linear systems.[98] More sophisticated molecular orbital arguments give us more information about the bending process, for example, how on electronic grounds the nitrosyl is expected to bend relative to the other ligands in the complex.

14 Rings, Cages, and Clusters

In the preceding chapters we have concentrated on coordination compounds both of main group and transition metal systems. Here the electronic controls on the structure originated from the MY or AY interactions. By way of contrast the factors holding together closed systems of atoms in the form of rings, clusters, or cages are associated with the direct YY interactions, and different considerations will apply. For example cubal coordination in a molecular system is very rare; $U(NCS)_8^{4-}$ in one of its salts with Et_4N^+ as counterion is an example, since ligand–ligand repulsions destabilize the structure. Yet it is the very nature of the close "ligand–ligand" approaches which allow the cage molecules based on the cube to hold together, for example, cubane C_8H_8, $[PhAlNAr]_4$, and $Pb_4(OH)_4^{4+}$ (Figure 14.1). In recent years considerable progress has been made in the generation of simple models[139,181,210–212,221] that allow prediction of cage geometry both in main group (particularly carboranes) and transition metal cluster compounds containing a variety of ligands bound to the metal. In the next chapter we see some of these ideas applied to the structures of extended caged arrays, namely, solid-state structures. There are basically two simple ways the structures of these molecules have been approached, and they are closely related to each other.

14.1 Molecular Orbital Structures of Simple Clusters

We initially consider the molecular orbital structure of the tetrahedral A_4 moiety[114] exemplified by the gaseous P_4 molecule. The orbitals for this unit have been derived in Section 9.1 where we were interested in the interaction of four tetrahedrally disposed ligands with a central atom. Here we stress the importance of "ligand–ligand" interactions. The de-

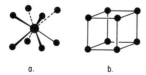

a.   b.

Figure 14.1 (*a*) Rare example of cubal eight-coordination in U(NCS)$_8^{4-}$ and (*b*) the structure of cubane C$_8$H$_8$. (Only the skeletal atoms of cubane are shown, and only the N atoms of the uranium complex.)

rivation of the molecular orbital diagram for A$_4$ is really quite simple since whether orbitals are bonding or antibonding depends upon their symmetry description (Figure 14.2). Since the *s* orbitals and radial *p* orbitals transform as identical symmetry species ($a_1 + t_2$) they mix together to produce two sets of hybrid orbitals, one inward pointing and the other outward pointing. The bonding or antibonding properties of the "π"-type (relative to a fictitious central-atom) orbitals may be seen by considering the signs of the overlap integrals between the atomic orbitals making up the molecular orbital. There result six orbitals to low energy involved in skeletal interactions, four more which point outward from the tetrahedron and may be used to attach the unit to another atom or hold a lone pair of electrons, and six higher-energy orbitals which are antibonding between the skeletal atoms. For P$_4$ each phosphorus atom contributes five valence electrons to this diagram. This is just sufficient to fill all the low-energy skeletal orbitals and the four outward pointing orbitals (now lone pairs) to give a stable, electron-precise, closed-shell structure (10 pairs of electrons). With four lone pairs this leaves one low-energy pair of electrons per PP linkage.

Overlap integrals:

a_1 σ $3S_{s\sigma}$
            $2S_{p\sigma} + S_{p\pi}$

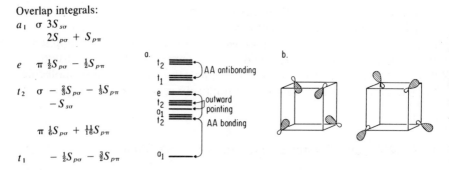

e π $\tfrac{1}{2}S_{p\sigma} - \tfrac{1}{2}S_{p\pi}$

t_2 σ $-\tfrac{2}{3}S_{p\sigma} - \tfrac{1}{3}S_{p\pi}$
            $-S_{s\sigma}$

        π $\tfrac{1}{6}S_{p\sigma} + \tfrac{11}{16}S_{p\pi}$

t_1 $-\tfrac{1}{2}S_{p\sigma} - \tfrac{3}{2}S_{p\pi}$

Figure 14.2 (*a*) Schematic orbital diagram for a tetrahedral A$_4$ unit. The overlap integrals for the various symmetry combinations are shown. $S_{s\sigma}$, $S_{p\sigma}$, and $S_{p\pi}$ are the overlap integrals between As orbitals, radial Ap orbitals, and tangential Ap orbitals, respectively. For the t_2 and a_1 orbitals the overlap integrals for the set of *s* orbitals and set of *p* orbitals are given separately. These will mix together to give "inward" and "outward" pointing hybrids shown in (*b*) for the orbitals of a_1 symmetry.

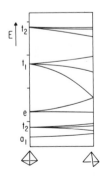

P_4 B_4H_{10} $P_4 cy_4$

S_2F_2 PCl_3 **Figure 14.3** Structures of some four-atomic molecules.

With two more electrons one of the skeletal antibonding orbitals has to be occupied, and this will weaken AA bonding. Since the t_1 orbital is delocalized over all four A atoms, a general loosening up of the structure might be expected; but with this electronic configuration the molecule is Jahn–Teller unstable and should distort. An example of a molecule with 11 skeletal pairs is B_4H_{10} (we see in the next section how to count up the number of skeletal pairs in these systems), and the overall effect has been (Figure 14.3) fission of one of the tetrahedral bonds. A similar effect holds in 12 electron-pair molecules (e.g., P_4(cyclohexyl)$_4$) with two tetrahedral bonds broken. With 13 pairs a basically tetrahedral structure is expected but with three linkages broken. One example is PCl_3 where the molecule has three short (P–Cl) and three long (Cl–Cl) distances. An alternative arrangement of the four atoms is the S_2F_2 molecule. With 14 pairs, one pair is located in the t_2 orbital and there are insufficient two electron-pair bonds to hold four atoms together. The PCl_2 molecule perturbed by an argon atom (a ''van der Waals molecule'') may be regarded as an example of this configuration. The presence of a central atom in the cage does nothing to affect the symmetry properties of these *skeletal* orbitals although it may add or remove electrons to or from the skeletal manifold. Thus CCl_4 is an interesting example of a 16-pair structure where there are no bonds between the skeletal Cl atoms. The molecule is of course held together by interaction with the central atom. That the ligand structure is a repulsive one if the carbon atom is neglected is readily seen if the skeletal Cl$^-$ ions are viewed as isoelectronic Ar atoms.

Figure 14.4 Energy changes on distortion of a tetrahedron by lengthening one of its edges. Only the highest five sets of orbitals of Figure 14.2 are shown. Note the stabilization afforded the lowest energy component of the t_1 orbital on distortion. This is the HOMO for the seven skeletal-pair molecule.

Figure 14.5 Structures of some eight-atomic molecules.

An alternative way to view the bond breaking process is to follow the energetic behavior of the occupied high energy orbitals on distortion. Figure 14.4 shows the energy changes of the molecular orbitals of the A_4 tetrahedron on increasing one of the AA distances. The driving force for the distortion of the seven skeletal pair system is the dramatic energy stabilization of the HOMO. The effect of the extra electrons is not always to break bonds. A change in angular geometry may occur instead. Thus PF_3 is pyramidal (13 pairs), but ClF_3 (14 pairs) is a T shape.

Another interesting series is the eight-atom family. Figure 14.5 shows a series of systems starting not with the 12-pair electron precise cubane structure but its sterically more favorable isomer cuneane. For the 14-, 15-, and 16-pair molecules, two, three and four bonds of the parent are broken.

Both As_4S_4 and S_4N_4 structures are found. The two different structures are partly understandable in that in As_4S_4 the more electronegative S atom prefers the low coordination site but in S_4N_4 the nitrogen atom is the more electronegative species and occupies the sites of lower coordination number. For S_8^{2+} and S_8 itself the endo–endo configuration is predicted. In fact the exo–endo and exo–exo conformations are found to be more stable in practice, but the isomer energy differences are probably small.

The approach has considered skeletal σ bonding only. If π bonding is important, then some of the skeletal electrons may be absorbed into a π network. For example the seven-pair S_4^{2+} species is planar (compared with B_4H_{10} of Figure 14.3) and has only four skeletal σ bonds. B_4H_{10} has five such linkages. Of the skeletal pairs, four form σ bonds while the remaining three occupy π orbitals. The system is thus isoelectronic with $C_4H_4^{2-}$. The six vertex example C_6H_6 has nine skeletal pairs. The prismatic structure (Figure 14.6) could accommodate these nicely to form an elec-

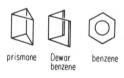

prismane Dewar   benzene
          benzene

Figure 14.6 Benzene isomers.

tron precise structure (prismane), but two σ bonds can be broken to give Dewar benzene with two π bonds. In cases where extra stabilization may arise through a delocalized π network, the simple rules based on the σ structure break down.

14.2 Electron Counting and Cage Geometry

This approach is based on a simple observation.[131-133] For the deltahedra shown in Figure 14.7, each n vertex polyhedron has $n + 1$ low-energy orbitals involved in skeletal bonding. Thus $n + 1$ skeletal electron pairs are sufficient to hold the framework together and give a stable structure. Since each n vertex deltahedron has $3n - 6$ edges (from Euler's theorem), these molecules are clearly electron deficient (if all vertices are occupied) in the true sense. For main group and transition metal systems there are simple rules for calculating the total number of skeletal electrons contributed by a collection of atoms without having to worry in detail

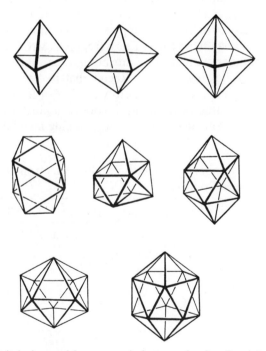

Figure 14.7 Deltahedra used by cage and cluster molecules. Reprinted with permission from *Advances in Inorganic Chemistry and Radiochemistry*, Vol. 19, Academic Press, New York. (1976), p. 1.

about molecular orbital arguments every time a new structure is considered. Main group elements may use three of their four valence shell orbitals (ns, np) for bonding within the cluster. See the directional nature of the orbitals of BH in Figure 13.9. If there is one singly bound terminal substituent (usually H) which may be a two- or one-electron ligand, then the fourth orbital is involved in bonding interactions with this species. If there is no substituent this orbital is allocated a lone pair of electrons. The number of electrons contributed by the skeletal atom to cluster bonding is then given by $s = v$ (two-electron ligand coordinated), $s = v - 1$ (one-electron ligand coordinated), or $s = v - 2$ (just a lone pair), respectively, where v is the number of valence electrons in the free atom. (If the skeletal atom has two substituents attached, e.g., AH_2, then the unit is regarded as AH^- (two-electron ligand coordinated) and the remaining H^+ is treated as if it were associated with the cluster. Having no electrons it cannot contribute anything to the electron count.

Table 14.1 shows the skeletal electron contributions for several species. For the carboranes, molecules of particular interest, a general formula holds for the number of skeletal pairs. For a molecule of general formula, $[(CH)_a(BH)_bH_c]^{d-}$, there are $p = \frac{1}{2}(3a + 2b + c + d)$ pairs. Transition metal atoms in cluster complexes often have coordinated CO or π bonded cyclopentadiene rings. For a fragment $M(CO)_n$ forming part of a cluster the number of electrons contributed to the skeletal framework is reached in the following fashion. Recall that in Section 13.2 it was suggested that for such a fragment there were $9 - n$ frontier orbitals. It is assumed that just three of these are involved in cluster bonding as in the main group case. This leaves $6 - n$ nonbonding metal orbitals which

Table 14.1 The Number of Skeletal Bonding Electrons that Main Group Cluster Units Contribute

Main Group Element A	A	Cluster Unit	
		AH or AY[a]	AH$_2$ or AY'
Li, Na	[−1]	0	1
Be, Mg, Zn, Cd, Hg	0	1	2
B, Al, Ga, In, Tl	1	2	3
C, Si, Ge, Sn, Pb	2	3	4
N, P, As, Sb, Bi	3	4	5
O, S, Se, Te	4	5	[6]
F, Cl, Br, I	5	[6]	[7]

From Ref. 212.
[a] Y = One-electron ligand; Y' = two-electron ligand.

are filled with electrons first. If there are a total of x electrons in the naked metal atom, then in the $M(CO)_n$ fragment there are $s = x - 2(6 - n)$ left over for cluster bonding. Table 14.2 shows some values for a selection of species containing terminal carbonyl groups and for the MC_p^- unit isolobal (Section 13.2) with $M(CO)_3$. Bridging CO groups are assumed to contribute two electrons to skeletal bonding. As with the treatment of the extra hydrogen atoms in the borane case it does not matter whether the CO groups are bound in a terminal or bridging fashion, since $M(CO)_n$ and $M(CO)_{n-1}$ plus a bridging carbonyl contribute overall the same number of skeletal electrons. Thus for a cluster species $M_a(CO)_b Cp_c^{d-}$ the number of skeletal electron pairs (p) is given by the formula $\frac{1}{2}[\Sigma v - 12a + 2b + 5c + d]$, where Σv is the total number of metal valence electrons, counting each metal as M^0.

Three sorts of geometries based on the deltahedra of Figure 14.7 are adopted by these species (Figure 14.8), and the structures may usually be readily predicted by the following approach known as Wade's rules. *Closo* structures are adopted by species where there are $n + 1$ skeletal pairs (s) and n skeletal atoms. There are just the correct number of atoms to occupy the n vertices of one of these deltahedra and just the right number of electron pairs to occupy all the low-lying orbitals. *Nido* species are those with $n + 2$ skeletal electron pairs but only n skeletal atoms. These n atoms occupy the n vertices of the $n + 1$ vertex deltahedron leaving one vacant. *Arachno* species are those with $n + 3$ skeletal pairs, n atoms, and here two sites of the $n + 2$ vertex deltahedron are vacant. For the $n + 4$ pair case the term *hypho* has been coined. In each case there are just the right number of electron pairs to occupy all the skeletal bonding orbitals. We first apply the scheme to the carboranes. Closo species are found for the series of compounds $C_a B_b H_{c+2}$ ($a = 0-2$). The

Table 14.2   The Number of Skeletal Bonding Electrons That Transition Metal Cluster Units Contribute

Transition Metal M	Cluster Unit			
	$M(CO)_2$	$M(\pi - C_5H_5)$	$M(CO)_3$	$M(CO)_4$
Cr, Mo, W	$[-2]$	-1	0	2
Mn, Te, Re	-1	0	1	3
Fe, Ru, Os	0	1	2	4
Co, Rh, Ir	1	2	3	5
Ni, Pd, Pt	2	3	4	6

From Ref. 212.

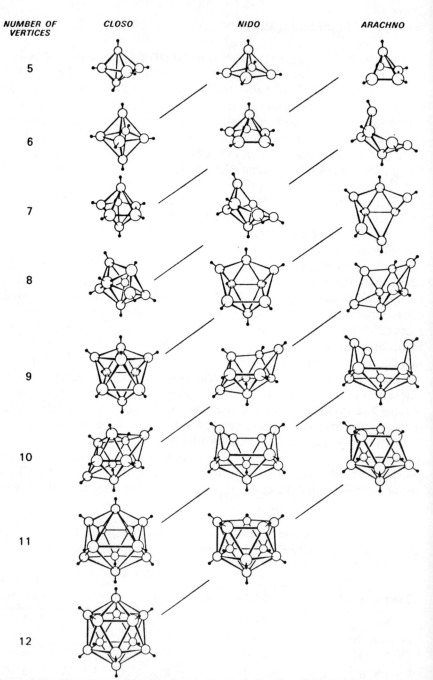

NUMBER OF VERTICES	CLOSO	NIDO	ARACHNO
5			
6			
7			
8			
9			
10			
11			
12			

Figure 14.8 Borane geometries showing structural relationships of closo, nido, and arachno species. Diagonal lines relate polyhedra with the same number of skeletal electron pairs but different numbers of skeletal atoms. Bridging hydrogens not shown. Reprinted with permission from R. W. Rudolph, *Acc. Chem. Res.* **9** 447 (1976). Copyright by the American Chemical Society.

$n = a + b$ skeletal atoms (C, B) occupy all the corners of the polyhedra of Figure 14.7 for $n = 5$–12. We find for all these molecules by doing the electron counting process that they have $n + 1$ pairs of skeletal electrons. Nido structures are found for compounds $C_aB_bH_{c+4}$ ($a = 0$–4), and arachno structures are adopted by compounds $C_aB_bH_{c+6}$. Table 14.3 summarizes data for these molecules. The molecules of the previous section may of course be viewed along similar lines. Thus P_4, with six skeletal pairs, will be based on a trigonal bipyramid but with one vertex missing—a nido structure (Figure 14.9). B_4H_{10}, with seven pairs, is an arachno structure based on an octahedron, and $P_4(cyclohexyl)_4$ is a hypho structure based on a pentagonal bipyramid. S_2F_2 and PCl_3 are more difficult to visualize. They should be "super-hypho" structures based on a square antiprism. The polyhedron that allows a better fit to their actual geometries is the cube. However as the number of empty vertices increases the flexibility which may occur in the rest of the structure away from the idealized geometry clearly increases and the applicability of the method diminishes. We saw similar deviations from the simple theory in the previous section for the S_8 and S_8^{2+} species.

The structures of many transition metal clusters with CO or πCp ligands fit in neatly on either of the two approaches. Thus $Rh_2Fe_2(\pi Cp)_2(CO)_8$ (Figure 14.10) has six skeletal pairs and has a nido trigonal bipyramid structure (cf. P_4). With seven pairs $Re_4(CO)_{16}^{2-}$ is isostructural with B_4H_{10}, an arachno geometry. $Mn_2(CO)_8Br_2$ with eight pairs has a hypho pentagonal bipyramidal structure [cf. $P_4(cyclohexyl)_4$].

$H_2Ru_6(CO)_{18}$ which may be regarded as formally derived from $Ru_6(CO)_{18}^{2-}$ has a total of seven skeletal pairs and its octahedral closo structure readily rationalized (Figure 14.11). Here there are fewer bonding electron pairs than metal–metal linkages, and the structure is certainly electron deficient. The loss of one carbonyl group (two electrons), the double negative charge, and then the addition of a carbon atom (four electrons) to the center of the cage produce the isoelectronic species $Ru_6(CO)_{17}C$ which of course is also octahedrally based. $Rh_6(CO)_{16}$, also with seven skeletal pairs, has a similar skeletal arrangement. $Fe(CO)_{15}C$ has seven skeletal pairs but has only five skeletal atoms. A nido octahedral structure (Figure 14.11) results.

Another interesting series of structures, in addition to the closo, nido, arachno, and hypho types, is found for several transition metal carbonyl clusters in which there are n skeletal atoms and n skeletal pairs. (All of

nido

arachno

hypho

Figure 14.9 Molecules of Figure 14.3 viewed using the deltahedra approach.

Table 14.3 Typical closo-, nido-, and arachno-Boranes and -Carboranes[a,b]

No. of Skeletal Bond Pairs	No. of Polyhedron Vertices	Fundamental Polyhedron	Closo Species $B_nH_n^{2-}$	Nido Species $B_nH_n^{4-}$	Arachno Species $B_nH_n^{6-}$
6	5	Trigonal bipyramid	$C_2B_3H_5$	—	$B_3H_8^-$
7	6	Octahedron	$B_6H_6^{2-}$ CB_5H_7 $C_2B_4H_6$	B_5H_9 $C_2B_3H_7$	B_4H_{10} —
8	7	Pentagonal bipyramid	$B_7H_7^{2-}$ $C_2B_5H_7$	B_6H_{10}, $B_6H_{11}^+$ $C_xB_{6-x}H_{10-x}$ $(x = 1 \to 4)$	B_5H_{11} —
9	8	Dodecahedron	$B_8H_8^{2-}$ $C_2B_6H_8$ $C_3B_5H_7$	— — —	B_6H_{12}

10	Tricapped trigonal prism	9	$B_9H_9^{2-}$ $C_2B_7H_9$	B_8H_{12} $C_2B_6H_{10}$	—
11	Bicapped Archimedean antiprism	10	$B_{10}H_{10}^{2-}$ $CB_9H_{10}^-$ $C_2B_8H_{10}$	$B_9H_{12}^-$ $C_2B_7H_{11}$ —	B_8H_{14} — —
12	Octadecahedron	11	$B_{11}H_{11}^{2-}$ $CB_{10}H_{11}^-$ $C_2B_9H_{11}$	$B_{10}H_{14}$ CB_9H_{13} $C_2B_8H_{12}$	B_9H_{15} $C_2B_7H_{13}$ —
13	Icosahedron	12	$B_{12}H_{12}^{2-}$ $CB_{11}H_{12}^-$ $C_2B_{10}H_{12}$	$CB_{10}H_{13}^-$ $C_2B_9H_{11}^-$ $C_4B_7H_{11}$	$B_{10}H_{15}^-$ $B_{10}H_{14}^{2-}$ —

From Ref. 212.

[a] The compound $C_4H_4B_6Me_6$ is not a nido carborane, but has an adamantane-type structure.

[b] See Figure 14.8.

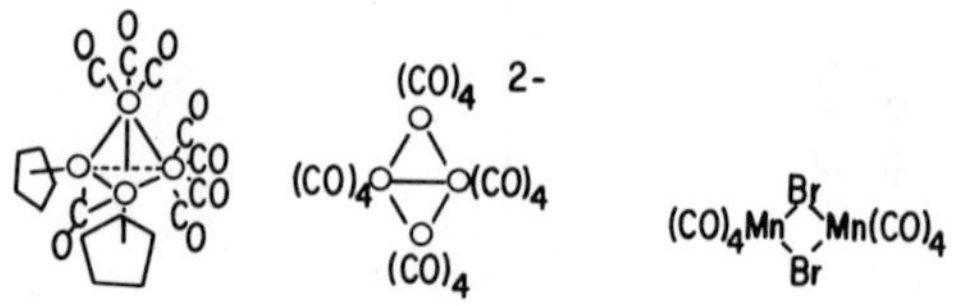

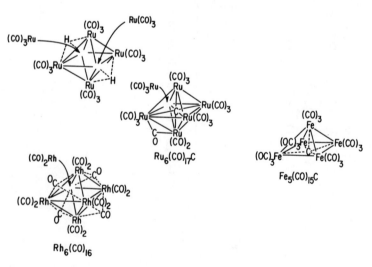

Figure 14.10   Some transition metal complexes with skeletal electron counts and structures analogous to those of the molecules of Figure 14.9: $Rh_2Fe_2Cp_2(CO)_8$ six pairs, $Re_4(CO)_{16}^{2-}$ seven pairs, $Mn_2(CO)_8Br_2$ eight pairs.

the structures just mentioned have an excess of electron pairs over skeletal atoms.) These adopt structures based on polyhedra with $n - 1$ vertices. The extra metal atom caps one of the triangular faces of the closo residue. In this position the three orbitals presented by the capping atom may interact with the three outward pointing orbitals of the triangular face without disturbing the skeletal bonding pattern of the rest of the cluster. This approach neatly accounts for several differences in geometry which are found for systems with the same number of metal atoms. Thus $Co_6(CO)_{14}^{4-}$ has seven skeletal pairs and an octahedral arrangement (closo octahedral) of Co atoms. $Os_6(CO)_{18}$ however, with six pairs, is a capped trigonal bipyramid (Figure 14.12). (This geometry is identical to a bicapped tetrahedron.)

Another example with seven pairs and seven atoms is the species $Rh_7(CO)_{16}^{3-}$ which analogously has a capped octahedral geometry. We use

Figure 14.11   Some transition metal cluster species with seven skeletal pairs.

Figure 14.12 Geometries of molecules with n skeletal pairs and n skeletal atoms, $Os_6(CO)_{18}$ and $Rh_7(CO)_{16}^{3-}$.

this second example to show how the capping process leads to the correct number of skeletal electrons associated with the octahedron. The capping group is in fact $Rh(CO)_{terminal}(CO)_{3\ bridging}$, leaving an $Rh(CO)_{12}^{3-}$ octahedral fragment with a total of nine skeletal electrons and all of the $6 - n$ nonbonding orbitals on each metal atom doubly occupied (Figure 14.13a). An $Rh(CO)_4$ fragment however contributes five electrons to the skeletal bonding of itself to the octahedron which occurs as we mentioned above between the three outward pointing orbitals of the octahedral face (one of these $6 - n$ orbitals on each atom) and the three skeletal orbitals of the $Rh(CO)_4$ unit. To aid our counting we take five electrons from these outward pointing hybrid orbitals, leaving one behind (Figure 14.13b), and fill the octahedral cluster bonding orbitals. There are then just enough (five) electrons in the skeletal bonding orbitals on the capping $Rh(CO)_4$ unit to fill all three octahedron-capping bonding orbitals (Figure 14.13c). Similar counting arguments may be made for any capping group and any deltahedral geometry.

In addition to holding for main group systems and transition metal cluster compounds, the approach is also very useful for looking at the structures of metalloboranes and metallocarboranes. Figure 14.14 shows some closo and nido systems. To understand the geometry we need only to combine the number of skeletal electrons contributed by the main group

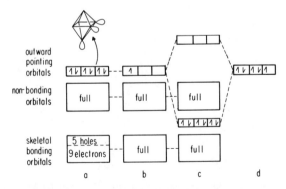

Figure 14.13 Electron counting procedure for n skeletal electron-pair–n skeletal atom structures. (This is a schematic diagram to help with the electron bookkeeping. We do not imply that the molecular orbital structure of this system is this simple.)

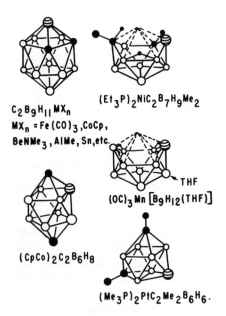

$C_2B_9H_{11}MX_n$

$MX_n = Fe(CO)_3, CoCp,$

$BeNMe_3, AlMe, Sn, etc.$

$(Et_3P)_2NiC_2B_7H_9Me_2$

$(OC)_3Mn[B_9H_{12}(THF)]$

↖THF

$(CpCo)_2C_2B_6H_8$

$(Me_3P)_2PtC_2Me_2B_6H_6.$

Figure 14.14 Some closo and nido metallocarboranes.

and transition metal-containing fragments together. Some interesting species occur when a metal atom is shared between two carborane polyhedra. Figure 14.15 shows two examples. The first structure may be regarded, as far as electron counting is concerned, as a complex with two nido $C_2B_9H_{11}^{2-}$ anions sandwiching the isoelectronic ions Fe^{2+}, Co^{3+}, and Ni^{4+}. These adopt symmetrical structures. With two more electrons "slipped" geometries are found. The structure of the $MC_2B_9H_{11}$ part of the molecule

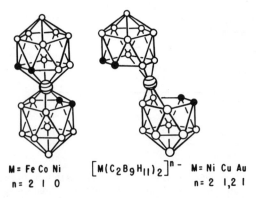

M = Fe Co Ni
n = 2 1 0

$[M(C_2B_9H_{11})_2]^{n-}$

M = Ni Cu Au
n = 2 1,2 1

Figure 14.15 "Slipping" of some carborane metal sandwiches.

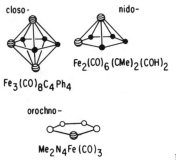

closo-

nido-

$Fe_3(CO)_8C_4Ph_4$

$Fe_2(CO)_6(CMe)_2(COH)_2$

orochno-

$Me_2N_4Fe(CO)_3$

⊖= metal   ●=carbon   ○=nitrogen

Figure 14.16 Organometallic compounds with eight skeletal pairs.

appears to be a nido structure based on the 13-vertex polyhedron which is needed to absorb the extra electrons. An alternative approach is to use the requirements of the 18-electron rule. The symmetrical sandwich species may be regarded as being isostructural with ferrocene (18 electrons). Two planar η^5 five-membered rings sandwich the Fe atom. With more metal electrons, in order to maintain a count of 18 electrons around the metal atom, the hapto number decreases, that is, the metal atom "slips" to the side of the ring, as mentioned in the previous chapter for some other apparently 20-electron Cp-containing species.

The approach is equally applicable to the structures of organometallic species. Figure 14.16 illustrates some species with eight skeletal pairs, and Figures 14.17 and 14.18 illustrate species with seven and six skeletal pairs. There are some exceptions to the scheme. The tetrahedral clusters

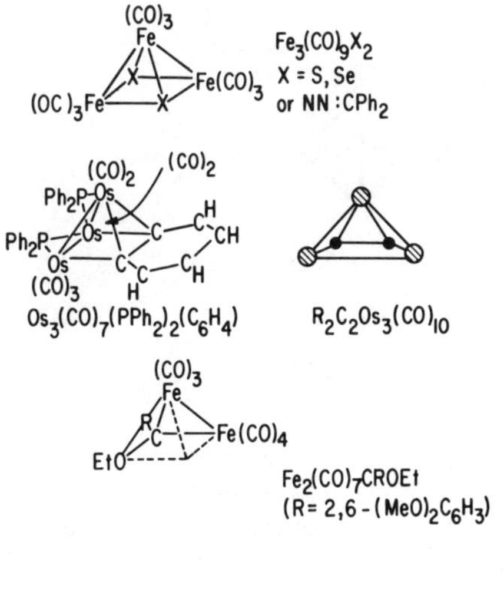

$(CO)_3$
Fe
$(OC)_3Fe$ —— X $Fe(CO)_3$

$Fe_3(CO)_9X_2$
X = S, Se
or NN :CPh_2

$(CO)_2$ $(CO)_2$
Ph_2P–Os
Ph_2P–Os—C
Os—C—C
$(CO)_3$
$Os_3(CO)_7(PPh_2)_2(C_6H_4)$

$R_2C_2Os_3(CO)_{10}$

$(CO)_3$
Fe
EtO —— Fe$(CO)_4$
$Fe_2(CO)_7CROEt$
$(R = 2,6-(MeO)_2C_6H_3)$

Figure 14.17 Organometallic compounds with seven skeletal pairs.

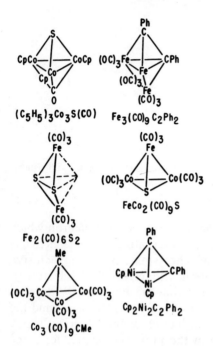

Figure 14.18 Organometallic compounds with six skeletal pairs.

$SCo_3(CO)_9$ and $BuNNi_3Cp_3$ have one electron too many, and the trigonal bipyramidal molecules $S_2Co_3Cp_3$ and $S_2Ni_3Cp_3$ have two and five electrons in excess of the number required.

Many metal hydrocarbon complexes may also be viewed in this way. Figure 14.19 shows some examples and underscores the fact that the fron-

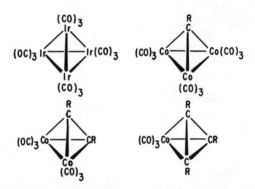

Figure 14.19 Metal hydrocarbon complexes showing the isolobality and isoelectronic nature of CH(CR) and $Co(CO)_3$ (tetrahedranes, C_4R_4, have recently been been isolated).

$Rh_8(CO)_{19}C$ **Figure 14.20** Structure of $Rh_8(CO)_{19}C$, an exception to the scheme.

tier orbitals of CH_2 are very similar (using the fragment formalism language of Chapter 13) to those of $Fe(CO)_4$ (see Tables 14.1 and 14.2), and CH is isolobal and isoelectronic with $Co(CO)_3$ [$Ir(CO)_3$]. In general the analogy between the behavior of main group and transition metal fragments is a very good one. As we have noted there are exceptions to the rules. One further exception is provided by the carbonyl cluster $Rh_8(CO)_{19}C$, which is expected on the scheme to be a closo dodecahedral species but is actually found (Figure 14.20) as a monocapped trigonal prism containing a central carbon atom with the remaining metal atom bridging one edge. There are also some ambiguities on the scheme. Thus four skeletal atom systems with a total of six skeletal pairs are predicted to have a nido trigonal bipyramid structure (Figure 14.18). There is a choice however of three or four coordinate vertices that may be left vacant. Leaving a four-coordinate vertex empty, as in $Fe_2(CO)_6S_2$, leads to a structure that is difficult to distinguish from one of the possible arachno octahedral structures (Figure 14.9).

The general idea however of $n + 1$ filled low-energy orbitals and n vertices may be used to roughly correlate the BB distances in some of these clusters.[212] Assuming an equal distribution of charge (which is not true as we have mentioned above), each vertex will be associated with $(n + 1)/n$ electron pairs. If these pairs are shared equally with all x linkages radiating from a particular x-coordinate vertex, then a measure of an edge bond order linking two atoms which are x_1 and x_2 coordinate might be the function $[(n + 1)/n][(1/x_1 + 1/x_2)]$. A plot against the observed bond lengths is shown in Figure 14.21.

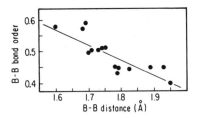

Figure 14.21 Plot of observed BB distances against the function $(n + 1)/n$ $(1/x_1 + 1/x_2)$, where n is the number of polyhedral vertices and the B atoms defining the linkage are x_1 and x_2 coordinate.

One final point to realize about the scheme is that the detailed molecular orbital structure of these transition metal complexes is usually much more complicated than we have suggested in this chapter. The simple approach described here, while working well in a large number of cases, is quite a simplification[140] of the molecular orbital picture found in detailed quantitative studies.

15 The Structures of Extended Arrays

In contrast to the structures of simple molecules there are a large number of possibilities for the geometric arrangement of simple systems as extended arrays in solids.[66,86,209] We may pinpoint three general factors that determine a solid-state structure[159,163]: (1) The efficient packing together of the atoms in the solid to give a dense high-symmetry arrangement. The occurrence of a particular AX structure will often be restricted if the radius ratio of the two atomic species A and X is unfavorable leaving atoms able to "rattle" in the structure. (2) The preservation of long-range order. Many structures look very favorable when a small cluster of atoms is built up, but sometimes completion of another coordination shell is not possible while retaining the symmetry of the structure. (3) Electronic forces. These may be divided into ionic nondirectional effects and covalent forces directional in character. It is with this last area that we are concerned. Rather than develop a set of rules to predict which structures compounds will exhibit in the solid state as extended arrays (a general theory for this immensely complex field has yet to be formulated anyway), we shall use the ideas of the previous chapter to understand some of the structural features of simple solids. We do however note the very interesting division of structural types[163–165,170] as a function of principal quantum number and electronegativity difference of Figure 15.1a and the XPS (X-ray photoelectron spectroscopy) results of Figure 15.1b.

15.1 Bond Breaking and the Number of Valence Electrons

A large number of AX systems with a total of eight valence electrons per formula unit adopt the zincblende (sphalerite) structure or the wurtzite structure (Figures 15.2 and 15.3). ZnS, carbon (diamond), AlSb, CuCl,

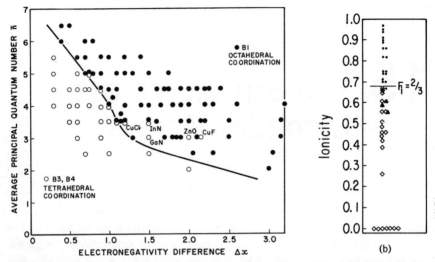

Figure 15.1 (*a*) Plot of electronegativity difference against principal quantum number for some AX species showing division into various structural types (from Ref. 163). (*b*) Observed "ionicity" parameter from XPS studies of the splitting of the most tightly bound two peaks in the XPS spectra of these binary systems. The critical value separating four- and six-coordinate structures occurs at an ionicity of ~2/3. Zincblende structure, diamonds; wurtzite structure, triangles; rocksalt structure, circles; CsCl structure, squares. Adapted from Ref. 118.

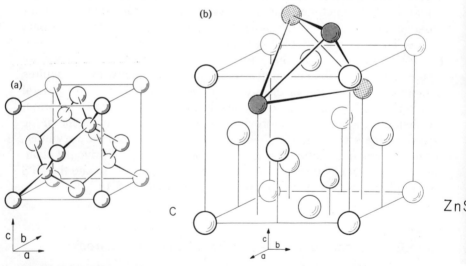

Figure 15.2 (*a*) The cubic diamond structure. (*b*) The derivative sphalerite or zincblende (ZnS) structure obtained by replacing carbon atoms of the diamond structure alternately with Zn and S atoms such that each atom is tetrahedrally coordinated by four atoms of the opposite type. Reprinted with permission from *The Crystal Chemistry and Physics of Metals and Alloys,* by W. B. Pearson, Wiley, New York (1972).

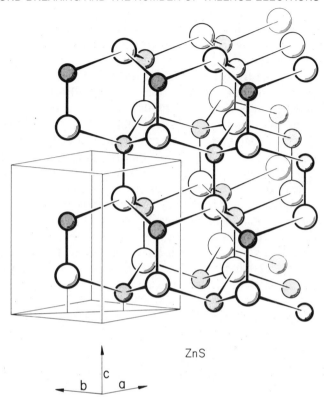

ZnS

Figure 15.3   Wurtzite (ZnS) structure. Both Zn and S are tetrahedrally coordinated by four atoms of the opposite type. The hexagonal unit cell is outlined. Reprinted with permission from *The Crystal Chemistry and Physics of Metals and Alloys,* by W. B. Pearson, Wiley,New York (1972).

CuBr, and HgS are among the systems that adopt the former structure, and ZnO, AlN, BeO, CdS, CuI, MgTe, and hexagonal diamond adopt the latter. Some of the examples are dimorphic. Extraction of a chunk of lattice from each structure reveals the basic building block (Figure 15.4) and demonstrates the similarity between the two systems. We recognize

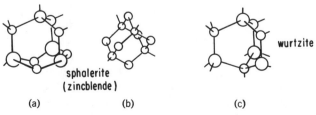

(a)   (b)   (c)

Figure 15.4   The building blocks of (*a*), (*b*) the zincblende, and (*c*) the wurtzite lattices.

in these two lattice fragments the framework found for adamantane or
P_4O_6 and P_4O_{10} (zincblende) or bicyclo(222)octane or barrellene (wurtzite)
in molecular systems.

How the molecule (or solid) is held together is best illustrated for the
case of the two diamond configurations. Each carbon atom has four
equivalent, tetrahedrally directed orbitals on a localized equivalent molecular orbital picture, made by suitable combination of 2s and 2p orbitals.
With four electrons, each carbon atom contributes half a bonding electron
pair to either a cage bond (i.e., toward the framework of the structures
of Figure 15.4) or to an outward pointing bond which connects this cage
to another. Thus no electrons go into antibonding orbitals, and there are
just the right number of electrons to make four bonds around each carbon
atom. Alternatively on a local basis each CY_4 unit has four pairs of electrons residing in bonding orbitals. In the case of molecular P_4O_6 the outward-pointing orbitals are filled with lone pairs of electrons leaving just
24 electrons (i.e., 12 pairs) to make all 12 intracage bonds. These are
examples of electron precise structures, and regular geometries are found.
But what happens when more than eight electrons per atom pair are present? According to the approach of Section 14.1, bonds are broken. The
scheme of Equation 15.1 shows one series of structures we shall follow
which show that the "molecular" approach also works well for solids:

$$ZnS \text{ (wurtzite)} \rightarrow GaSe \rightarrow As \rightarrow Se, Te \rightarrow I_2 \rightarrow Xe$$

$$(15.1)$$

No. of electrons   8   9   10   12   14   16

We start off by considering the wurtzite structure. Each cage (Figure
15.3) contains two formula units. The three AX units forming the sides
of the prism are shared by three such cages, the "apical" AX unit is
associated with only one cage. The apical atoms are attached to another
cage by a single linkage top and bottom. The prismatic atoms are each
attached to two other cages. There are thus 16 electrons per cage, all
residing in bonding orbitals. With one extra electron per formula unit (for
example GaSe) there are sufficient electrons to break one two-electron
bond per cage. We need to count up the number and type of the bonds
present in the cage very carefully in order to see how many bonds may
be broken. We may break either two extra-cage linkages (a) in Figure
15.5, three prismatic cage linkages (b), or two pyramidal (c) linkages with
these electrons. If extra-cage bonds are broken then each cage will contribute to the bond breaking, and two of these bonds may be broken for
each pair of electrons contributed by a single cage. If cage bonds themselves are broken, the number which may be broken per extra electron

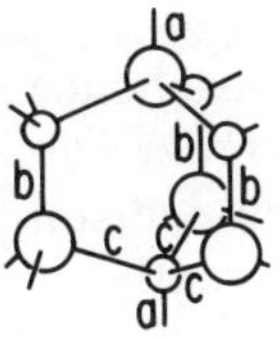

Figure 15.5 Types of linkages in the wurtzite cage.

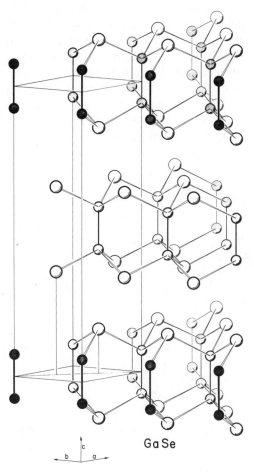

GaSe

Figure 15.6 Structure of β-GaSe (small circles, Ga). Reprinted with permission from *The Crystal Chemistry and Physics of Metals and Alloys,* by W. B. Pearson, Wiley, New York (1972).

pair per cage depends on the number of cages that share a particular linkage: three for the prismatic bonds, b, and two for the pyramidal bonds, c. The structure of β-GaSe (Figure 15.6) shows that the bonds broken are either both the a bonds or all three b bonds, the result being indistinguishable as a result of the symmetry of the structure. Importantly the number and type of linkages broken compared to the wurtzite structure are in accord with the number of extra electrons present. Two other differences are of note by comparison with the wurtzite geometry. First after bond fission, each of the layers moves with respect to the one beneath it. This clearly aids the packing together of the layers. Second a readjustment of the atom positions within the cage has occurred. The Se atoms only occupy the apical positions of the wurtzite cage fragment.

A structure where different linkages of the wurtzite cage are broken is that of the mineral wolfsbergite, $CuSbS_2$. If we write this as $Cu^ISb^{III}S_2$ to satisfy divalent sulfur, then the copper contributes one $4s$ electron to the electron count. Sb contributes five electrons, and this leads on average to nine electrons per AS unit, that is, one more than in ZnS, and makes this structure isoelectronic with GaSe. Wolfsbergite is a superstructure of wurtzite, and we can readily see from Figure 15.7 that two pyramidal

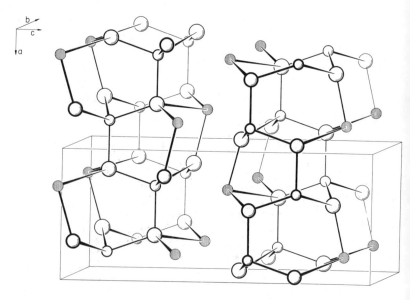

$CuSbS_2$

Figure 15.7 The structure of wolfsbergite, $CuSbS_2$. Large circles, S, small open circles, Cu. Reprinted with permission from *The Crystal Chemistry and Physics of Metals and Alloys*, by W. B. Pearson, Wiley, New York (1972).

SbS linkages are broken. Again the sites of lowest coordination, where the breakage has occurred, contain the most electronegative species. This observation is in accord with a general feature of molecular compounds. The sites of lowest coordination are invariably those carrying the highest charge. Just as in our discussion of the forces determining ligand site preferences in main group compounds of Section 7.1, the most electronegative ligand will prefer the sites of highest electron density. (Recall that ClF_3 contains the most electronegative ligands (F) in the sites of lowest (one) coordination number.)

In enargite, Cu_3AsS_4 ($Cu^I Cu_2^{II} As^{III} S_4$), there is half an electron on average per AS formula unit. This is enough to break only one extra cage bond of the wurtzite unit but appears to be insufficient to produce a separation between two halves of the structure as in $CuSbS_2$ or GaSe (we need to break two bonds at least), and the result is a slightly distorted structure based on wurtzite, in which the atoms are displaced from their ideal positions. Instead of regular tetrahedral coordination a range of Cu–S and As–S bond lengths are found.

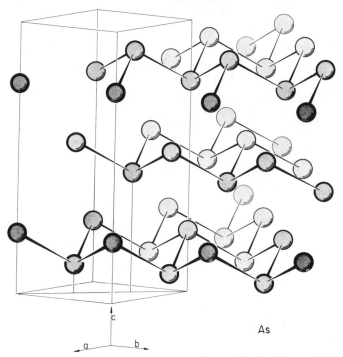

As

Figure 15.8 Layer structure of elemental arsenic. Reprinted with permission from *The Crystal Chemistry and Physics of Metals and Alloys*, by W. B. Pearson, Wiley, New York, (1972).

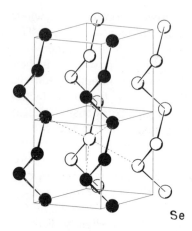

Se

Figure 15.9 Chain structure of Se and Te. Reprinted with permission from *The Crystal Chemistry and Physics of Metals and Alloys,* by W. B. Pearson, Wiley, New York (1972).

With 10 electrons per formula unit there are now sufficient extra electrons to break two bonds per cage. In addition to the two extra-cage linkages broken in GaSe, the three vertical linkages, *b*, may also be broken. (Recall each vertical cage linkage is shared by three cages, so that one extra pair of electrons contributed per cage is sufficient for fission of all three.) The As structure is just this, and again the layers are moved with respect to each other to aid in packing (Figure 15.8).

With 12 electrons one more bond may be broken per cage, and the result is the chain structure (Figure 15.9) found for elemental Se or Te. With 14 electrons another bond may be broken, and the result is a lattice composed of molecular I_2 units still retaining (Figure 15.10) in fact a geometric resemblance, via packing of the diatomic units, to the parent structures. The almost trivial case of Xe with 16 electrons has no bonds between the atoms, and a van der Waals solid results.

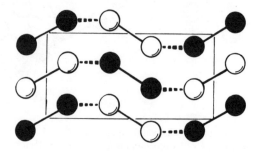

Figure 15.10 Molecular lattice of I_2.

Extra electrons may be introduced in other ways such as inclusion of an alkali or alkaline earth metal into the structure. Thus $CaSi_2$ (10 electrons per Si_2 unit) has a rumpled sheet structure just like As, and CaSi (12 electrons per Si_2 unit) has zigzag chains just like Se or Te (Figure 15.11). A similar analysis applies to silicate structures. SiO_2 itself exists in several modifications, the β-cristobalite and tridymite arrangements being closely related (via an oxygen atom spacer) to those of zincblende and wurtzite. It is also found, with Al^{3+} substituted for Si and a cation present to maintain electrical neutrality, as feldspars and zeolites. As the number of electrons per atom increases, this three-dimensional structure is broken up, leading eventually to orthosilicates containing isolated SiO_4^{4-} tetrahedra. Table 15.1 shows the trends in structural features found. A similar series of considerations apply to a wide range of sulfides and sulfosalts.

An alternative way of reacting to an increase in the number of electrons is via an angular distortion of the coordination geometry. Thus the four valence-pair molecule CF_4 is a regular tetrahedron, but SF_4 with two more electrons has distorted to its characteristic butterfly shape. Examination of the structures of some complex sulfur minerals containing electrons in an excess of the number necessary for regular tetrahedral coordination do in fact often show geometries of this type.[154]

Table 15.1 Breakup of silcate structures with increasing electron count.

System	Number of Electrons per Atom	Examples
SiO_2	5.33	Three-dimensional structures (with isomorphous replacement by Al, feldspars, and zeolites are found)
$(Si_2O_5)_n^{2n-}$	5.71	Double chains, e.g., gillespite ($BaFeSi_4O_{10}$), and sheets, e.g., micas
$(Si_4O_{11})_n^{6n-}$	5.87	Double chains, e.g., amphiboles and tremolite $[(OH)_2Ca_2Mg_5(Si_4O_{11})_2]$
$(SiO_3)_n^{2n-}$	6.00	Cyclic chains, e.g., benitoite ($BaTiSi_3O_9$), and linear chains, e.g., pyroxenes [$CaMg(SiO_3)_2$], and diopside
$(SiO_4)^{4-}$	6.4	Isolated tetrahedra in orthosilicates, e.g., olivines M_2SiO_4 (M Mixture Mg, Fe, Mn)

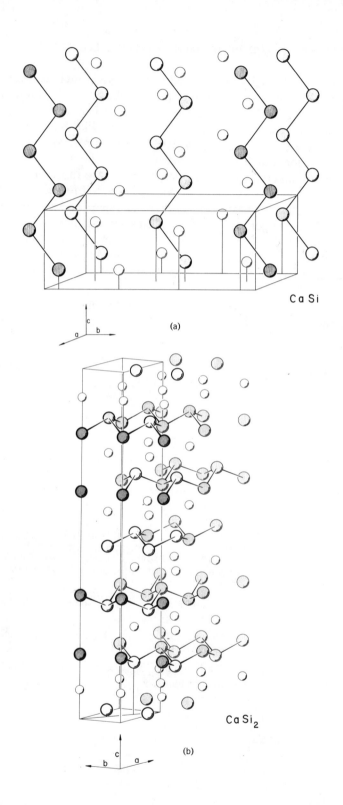

Ca Si

(a)

Ca Si₂

(b)

15.2 Electron Counting Schemes

A survey of solid-state structures based on the zincblende ZnS structure immediately shows some striking similarities to the molecular scheme of Section 14.2. The $CdIn_2Se_4$ and $CdAl_2S_4$ species crystallize as defect zincblende structures (Figure 15.12). We may represent them as $Cd\square In_2Se_4$ and $Cd\square Al_2S_4$, where $\square$ = vacancy to emphasize their stoichiometric relationship to ZnS. $ZnAl_2S_4$ (high-temperature form) has an analogous defect wurtzite structure. On average per AS or ASe unit (A = Cd, In, Al, or $\square$) there are eight electrons, just as in ZnS and an average of 12 electrons per "zincblende" cage. This cage structure is neatly regarded as a nido ZnS structure. The zincblende structure of Figure 15.2 would of course be regarded as a closo structure. Examination of a model of this system shows in fact that there are three different zincblende "cages." One of these has no atoms missing, the second has two missing, and the third has one missing. So the structure contains closo, nido, and arachno cages. Since there needs to be an average of four electrons per site, we

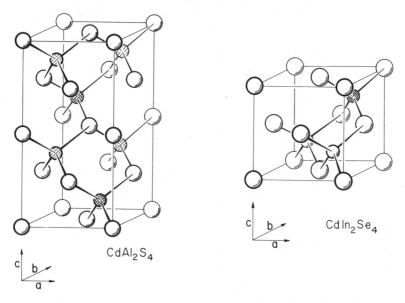

Figure 15.12 The defect zincblende structures of $CdIn_2Se_4$ and $CdAl_2S_4$ (shaded atoms, In or Al; crosshatched atoms, Cd). Reprinted with permission from *The Crystal Chemistry and Physics of Metals and Alloys*, by W. B. Pearson, Wiley, New York, (1972).

Figure 15.11 (a) Layer structure of $CaSi_2$ (cf. As) and (b) chain structure of CaSi (cf. Se, Te) (connected atoms are Si). Reprinted with permission from *The Crystal Chemistry and Physics of Metals and Alloys*, by W. B. Pearson, Wiley, New York, (1972).

Table 15.2 Solid-State Tetrahedrally Based Structures

Composition Formula	Chemical Formula
4 and 44	C, Si, Ge, Sn, SiC
35	BN, BP, BAs, AlN, AlP, AlAs, AlSb, GaN, GaP, GaAs, GaSb, InN, InP, InAs, InSb
26	BeO, BeS, BeSe, BeTe, BePo, MgTe, ZnO, ZnS, ZnSe, ZnTe, ZnPo, CdS, CdSe, CdTe, CdPo, HgS, HgSe, HgTe, MnS, MnSe
17	CuCl, CuBr, CuI, AgI, NH_4F, also CuH and CuD
25_2	ZnP_2, $ZnAs_2$, CdP_2
3_26	Al_2O?
136_2	$CuBSe_2$, $CuAlS_2$, $CuAlSe_2$, $CuAlTe_2$, $CuGaS_2$, $CuGaSe_2$, $CuGaTe_2$, $CuInS_2$, $CuInTe_2$, $CuTlS_2$, $CuTlSe_2$, $AgAlS_2$, $AgAlSe_2$, $AgAlTe_2$, $AgGaS_2$, $AgGaSe_2$, $AgGaTe_2$, $AgInS_2$, $AgInSe_2$, $AgInTe_2$, $CuFeS_2$, $CuFeSe_2$, $AgFeS_2$
245_2	$BeSiN_2$, $MgGeP_2$, $ZnSiP_2$, $ZnSiAs_2$, $ZnGeP_2$, $ZnGeAs_2$, $ZnSnAs_2$, $CdGeP_2$, $CdGeAs_2$, $CdSnAs_2$
1_356_4	Li_3PO_4, Li_3AsO_4, Cu_3PS_4, Cu_3AsS_4, Cu_3AsSe_4, Cu_3SbS_4, Cu_3SbSe_4
14_25_3	$CuSi_2P_3$, $CuGe_2P_3$, $CuGe_2As_3$
1_246_3	Cu_2SiS_3, Cu_2SiTe_3, Cu_2GeS_3, Cu_2GeSe_3, Cu_2GeTe_3, Cu_2SnS_3, Cu_2SnSe_3, Cu_2SnTe_3
156	CuAsS
3_246	Al_2CO
1_2246_4	Cu_2ZnGeS_4, Cu_2ZnSnS_4, Cu_2CdSnS_4, Cu_2HgSnS_4, Cu_2FeGeS_4, Cu_2FeSnS_4, $Cu_2FeGeSe_4$, $Cu_2FeSnSe_4$
12_236_4	$AgCd_2InTe_4$, $AgHg_2InTe_4$
134_25_4	$CuGaGe_2P_4$
23_245_4	$ZnIn_2GeAs_4$, $ZnIn_2SnAs_4$, $CdIn_2GeAs_4$ (part of a series of solid solutions between 245_2 and 35)
$2_53_206_8$	$Hg_5Ga_2Te_8$, $Hg_5In_2Te_8$
$12_33_306_8$	$AgHg_3In_3Te_8$
$1_22_2067_4$	$Ag_2Hg_2SI_4$
3_406_4	Ga_2S_2, Ga_2Se_2, Ga_2Te_2?, In_2Se_2
$1_25_206_4$	$CuSbS_2$, $CuSbSe_2$?, $CuBiS_2$
4_405_4	SiAs?, GeAs?
4_305_4	Si_3N_4, Ge_3N_4
2_2406_4	Be_2SiO_4, Be_2GeO_4, Zn_2SiO_4, Zn_2GeO_4
23_206_4	$ZnAl_2S_4$, $ZnAl_2Se_4$, $ZnAl_2Te_4$, $ZnGa_2S_4$, $ZnGa_2Se_4$, $ZnGa_2Te_4$, $ZnIn_2Se_4$, $ZnIn_2Te_4$, $CdAl_2S_4$, $CdAl_2Se_4$, $CdAl_2Te_4$, $CdGa_2S_4$, $CdGa_2Se_4$, $CdGa_2Te_4$, $CdIn_2Se_4$, $CdIn_2Te_4$, $HgAl_2S_4$, $HgAl_2Se_4$, $HgAl_2Te_4$, $HgGa_2S_4$, $HgGa_2Se_4$, $HgGa_2Te_4$, $HgIn_2Se_4$, $HgIn_2Te_4$

Table 15.2 (*Continued*)

Composition Formula	Chemical Formula
$1_2 2 0 7_4$	Li_2BeF_4, Cu_2HgI_4, Ag_2HgI_4
$2_3 0 5 7_3$	Zn_3PI_3, Zn_3AsI_3
$3_3 0 5 6_3$	Ga_3AsSe_3, In_3AsTe_3 (part of a series of solid solutions between $3_2 0 6_3$ and 35)
$1 3_2 0 6_3 7$	$CuIn_2Se_3Br$, $CuIn_2Se_3I$, $AgIn_2Se_3I$
$1 2 5 0 6_4$	$LiZnAsO_4$, $LiZnVO_4$
$1 3 4 0 6_4$	$LiAlSiO_4$, $LiAlGeO_4$, $LiGaGeO_4$
$1_2 6 0 6_4$	Li_2SO_4, Li_2SeO_4, Li_2MoO_4, Li_2WO_4
$4_2 0 5_4$	$GeAs_2$
$1_2 3_0 2_0 6 7_6$	$Ag_2Hg_3SI_6$
$3_2 0 6_3$	Al_2S_3, Al_2Se_3, Ga_2S_3, Ga_2Se_3, Ga_2Te_3, In_2Se_3, In_2Te_3
$0 5_4$	P, As, Sb, Bi
$4_2 0 6_2$	GeS, GeSe, SnS, SnSe
$4 0 6_2$	SiO_2, SiS_2, $SiSe_2$, GeO_2, GeS_2, $GeSe_2$?
$2 0 7_2$	BeF_2, $BeCl_2$, $BeBr_2$, BeI_2, $ZnCl_2$, $ZnBr_2$, ZnI_2, HgI_2
$3 5 0_2 6_4$	BPO_4, BPS_4, $BAsO_4$, $AlPO_4$, $AlPS_4$, $AlAsO_4$, $GaPO_4$, $GaAsO_4$, $FePO_4$, $MnPO_4$
$4_4 0_3 7_4$	CF
$5_2 0_2 6_5$	P_4O_6
$5_2 0_3 6_5$	P_4O_{10}
$0 6_2$	S, Se, Te
$3 0_2 7_3$	Al_2Cl_6, Al_2Br_6, Al_2I_6, Ga_2Cl_6, Ga_2Br_6, In_2Cl_6, In_2Br_6, In_2I_6
$4_2 0_3 7_4$	nCF_2 (polyfluorethylene)
$4_4 0_9 7_{12}$	C_2Cl_6, Si_2Cl_6
$5_4 0_7 7_8$	P_2I_4
$4 0_3 7_4$	SiF_4, SiI_4, GeI_4, SnI_4
$5_2 0_5 7_6$	PI_3, SbF_3, $SbCl_3$
$6_4 0_5 7_4$	S_2Cl_2
$6 0_2 7_2$	SCl_2
$0_3 7_4$	Cl_2, Br_2, I_2

Adapted from Ref. 158.

may write the formula of such a system as $0_a 1_b 2_c \ldots N_n \ldots 7_h$ where n atoms per formula unit each contribute N_n electrons. Clearly a–h must be chosen so that $\Sigma_n n N_n / \Sigma_n n = 4$. This is the Grimm–Sommerfeld valence rule. An entry 0 corresponds to a defect structure. Some examples for allowed values of a–h are shown in Table 15.2. It is interesting to note

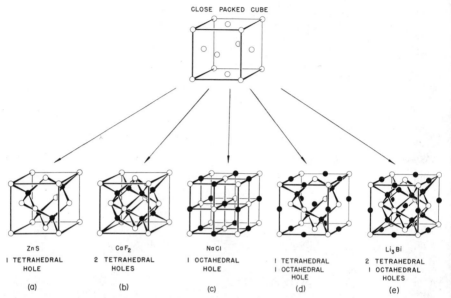

Figure 15.13 Schematic derivation of filled up derivatives from a simple structure. Reprinted with permission from *The Crystal Chemistry and Physics of Metals and Alloys*, by W. B. Pearson, Wiley, New York (1972).

for example that the layer structure of elemental arsenic may be regarded as a defect wurtzite structure on this scheme.

Solid-state chemists have used electron counting ideas similar to these for many years,[158] which may be formulated as four principles:

1. Starting with a close-packed cubic array of "anions," filling of the various holes gives the structures shown in Figure 15.13.

2. By substituting in an ordered fashion different cations with the same or average number of electrons then a superstructure may be built up.

3. By substituting cations with more electrons per atom some of the sites of the structure remain vacant in an ordered fashion.

4. By substituting cations with less electrons per atom, filled-up derivatives are obtained.

Rule 1 is the solid-state analog of the presentation in Figure 14.9 of the fundamental polyhedra that we consider. Rule 2 has no molecular analog in the strictest sense since superstructures are essentially a solid-state concept but the best analogy is that similar structures with the same number of skeletal atoms are obtained on substitution of for example a BH group by a $Co(CO)_3$ group (for example the isoelectronic and iso- structural molecules of Figure 14.14). Rule 3 recalls the occurrence of closo, nido, and arachno molecular species. Rule 4 has its molecular coun- terpart in cluster compounds with a central atom, for example $Ru_6(CO)_{17}C$ of Figure 14.11. The analogy between the two schemes applied to mo- lecular species and extended arrays is then a very striking one.

15.3 The Fragment Formalism

As an example of the use of the fragment formalism in solid-state chemistry we examine the structures of some tetrahedrally based solids.[30] The wurtzite and zincblende structures of the previous sections may be regarded as being assembled from puckered arsenic-like sheets of Figure 15.8. Depending upon whether the adjacent sheets are eclipsed or stag- gered with respect to each other (Figure 15.14), the wurtzite or zincblende structures arise respectively. We shall not be concerned here with the factors stabilizing one arrangement over another but inquire what are the electronic requirements for AA, XX, and AX linkages between these puckered sheets. The ZnS structures we have just mentioned with eight electrons per formula unit of course contain AX linkages, but the GaSe (nine electrons) structure of Figure 15.6 exhibits AA linkages, and a part of the more complex covellite structure which contains Cu^IS (seven elec- trons) shows XX linkages in the form of S_2 pairs. The obvious choice of fragments with which to build up the structure is a pair of infinite two- dimensional sheets. However we shall be able to get just as much infor- mation by looking at the orbitals of a puckered regular hexagon of atoms terminated with H atoms to simulate its attachment to the other atoms

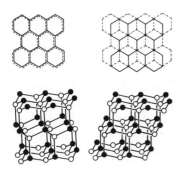

a.   b.

Figure 15.14 Generation of wurtzite and zincblende structures by stacking of puckered sheets.

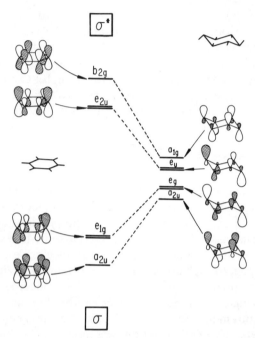

Figure 15.15 Change in energy and character of benzene π-type orbitals on puckering.

of the sheet—namely a distorted benzene molecule. Figure 15.15 shows how the frontier orbitals of the benzene molecule (the easily visualized π-type orbitals) change in energy as the geometry is changed. Figure 15.16 shows how their description changes when atoms of different electronegativity are introduced into the ring. The lowest-energy orbitals are primarily associated with the more electronegative species (see Section 1.3) and the higher-energy orbitals with the least electronegative atoms.

Figure 15.17 shows schematic molecular orbital diagrams for interactions of the three types derived by allowing the frontier orbitals of each sheet to interact with those on an adjacent sheet. Enclosed in blocks are the three orbitals derived from the π and π^* orbitals of the planar fragment. Thus each rectangular block of orbitals contains one a_1 and one e species orbital, and each will be filled with a total of six electrons. In the AA-linked case, because the lower-energy set of outward-pointing orbitals on one ring are associated largely with the orbitals on the X atoms and point away from the adjacent ring, the energy changes associated with them are small. These lead to the orbitals labeled I and II. The largest energy changes are associated with the higher-energy set of orbitals, largely A located since here overlap between orbitals on adjacent rings

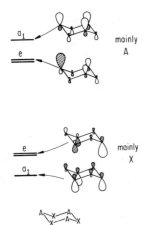

Figure 15.16 Frontier orbitals of $A_3X_3H_6$ units.

is largest. The result is a set of three AA bonding orbitals (III) and a set of three AA antibonding orbitals (IV).

The converse is true for the case where the two sheets are linked by the X atoms. Here the largely X-located orbitals lead to the largest interactions to give XX bonding (I) and antibonding (II) trios of orbitals. The higher-energy (largely A-located) orbitals point away from the adjacent sheet and remain approximately intersheet nonbonding.

For the third case where AX linkages occur then the largest interactions are between high-energy orbitals on one sheet (largely A-localized) with low-energy orbitals on the other (largely X-localized). The remaining low-energy orbital on one sheet to interact with a high-energy orbital on the next. Using the criterion that electrons in strongly antibonding orbitals lead to structural instability we can readily see that the AA-linked structure is favored for nine electrons per formula unit. (With n electrons per AX formula unit the "benzene" molecule of Figure 15.15 has the configuration π^0, π^3, π^6, $\pi^6\pi^{*3}$, $\pi^6\pi^{*6}$ for $n = 6, 7, 8, 9$, and 10.) This leads to an electron configuration $I^6II^6III^6$. The AX-linked arrangement is favored for eight electrons with the configuration I^6II^6 and the XX-linked arrangement for seven electrons per formula unit to give the configuration I^6.

In agreement with these ideas, such structural features are indeed observed where predicted. In both wurtzite and zincblende modifications of ZnS and in all other AX systems with this structure and superstructures derived from them, all of which have eight electrons per formula unit, AX linkages are found. Thus AlSb, CuI, and LiGaO$_2$ contain AlSb$_4$, CuI$_4$, and LiO$_4$ and GaO$_4$ tetrahedra, respectively. In the nine-electron GaS and

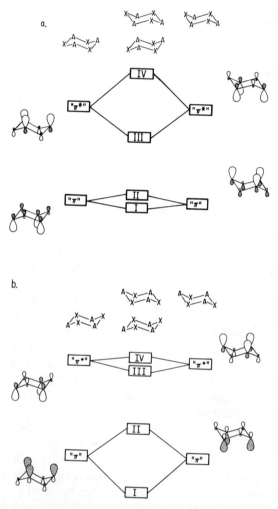

Figure 15.17 Schematic molecular orbital diagrams obtained by linking adjacent puckered sheets with (*a*) AA, (*b*) XX, and (*c*) AX contacts.

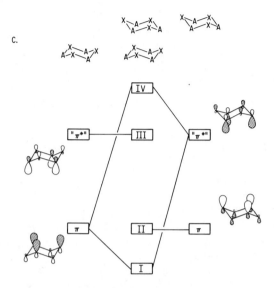

Figure 15.17   (Continued)

GaSe structures (Figure 15.16), AA linkages are found. In the seven-elec-
tron Cu^IS region of the complex covellite ($Cu_2^ICu^{II}S_3$) structure, sulfur-
linked (XX) sheets are found. With 10 electrons per atom pair the electron
configuration is $I^6II^6III^6IV^6$, and none of the three modes of attachment
gives rise to bound pairs of sheets. Thus the arsenic structure consists
of isolated puckered sheets. These results are good evidence for covalent
directional interaction between the atoms, that is, electronic control of
the geometry. While the results of this chapter do give a rather interesting
insight into solid-state structures, we are still a long way off being able
to rationalize or predict solid-state structures *a priori*. This is a challenge
which structural chemists must take up in the future.

References

1. D. M. Adams, *Metal Ligand and Related Vibrations*, St. Martins Press, New York (1968).
2. C. J. Adams and A. J. Downs, *Chem. Commun.* 1699 (1970).
3. W. J. Adams, H. B. Thompson, and L. S. Bartell, *J. Chem. Phys.* **53**, 4040 (1970).
4. T. A. Albright, *A Holiday Coloring Book of Fragment Molecular Orbitals*, Cornell University (1977), unpublished.
5. R. F. W. Bader, *Mol. Phys.* **3**, 137 (1960).
6. C. J. Ballhausen and H. B. Gray, *Molecular Orbital Theory*, Benjamin, New York (1964).
7. C. J. Ballhausen, in Ref. 191, part B, p. 129.
8. C. J. Ballhausen, *Introduction to Ligand Field Theory*, McGraw-Hill, New York (1962).
9. V. Balzani and V. Carassiti, *Photochemistry of Coordination Compounds*, Academic Press, New York (1970).
10. L. S. Bartell, *J. Chem. Educ.* **45**, 754 (1968).
11. L. S. Bartell, *J. Chem. Phys.* **32**, 827 (1960).
12. L. S. Bartell and C. J. Marsden, *J. Chem. Soc.* (Dalton) 1582 (1977).
13. L. S. Bartell and V. Plato, *J. Am. Chem. Soc.* **95**, 3097 (1973).
14. H. Basch, A. Viste, and H. B. Gray, *Theoret. Chim. Acta* **3**, 458 (1965).
15. A. Begum, S. Subramanian, and M. C. R. Symons, *J. Chem. Soc.* (A) 918 (1970).
16. S. Bellard, K. A. Rubinson, and G. M. Sheldrick, *Acta Crystallogr.* **B35**, 271 (1979).
17. R. S. Berry, *J. Chem. Phys.* **32**, 933 (1960).
18. H. Bethe, *Ann. Phys.* **3**, 135 (1929).
19. BIDICS, *Bond Index Determination of Inorganic Crystal Structures*, I. D. Brown Institute for Materials Research, McMaster University, Hamilton, Ontario, Canada (1969-).
20. R. J. Buenker and S. D. Peyerimhoff, *Chem. Rev.* **74**, 127 (1974).
21. D. C. Bradley, *Chem. Brit.* **11**, 393 (1975).
22. R. D. Burbank and G. R. Jones, *J. Am. Chem. Soc.* **96**, 43 (1974).
23. J. K. Burdett, *Coord. Chem. Rev.* **27**, 1 (1978).
24. J. K. Burdett, *J. Chem. Soc.* (Faraday II) **70**, 1599 (1974).
25. J. K. Burdett, *Adv. Inorg. Chem. Radiochem.* **21**, 113 (1978).
26. J. K. Burdett, J. M. Grzybowski, R. N. Perutz, M. Poliakoff, J. J. Turner, and R. F. Turner, *Inorg. Chem.* **17**, 147 (1978).

27. J. K. Burdett, R. Hoffmann, and R. C. Fay, *Inorg. Chem.* **17**, 2553 (1978).

28. J. K. Burdett, *Chem. Soc. Rev.* **7**, 507 (1978).

29. J. K. Burdett, *J. Am. Chem. Soc.* **101**, 580 (1979).

30. J. K. Burdett, *J. Am. Chem. Soc.* **102**, 450 (1980).

31. J. K. Burdett and T. A. Albright, *Inorg. Chem.* **18**, 2112 (1979).

32. H. B. Bürgi, *Angew. Chem.* (Int. Ed.) **14**, 460 (1975).

33. G. Burns, *J. Chem. Phys.* **41**, 1521 (1964).

34. R. G. Burns, *Mineralogical Applications of Crystal Field Theory,* Cambridge University Press, Cambridge (1970).

35. V. Calder, D. E. Mann, K. S. Seshadri, M. Allavena, and D. White, *J. Chem. Phys.* **51**, 2093 (1969).

36. J. Chatt and L. A. Duncanson, *J. Chem. Soc.* 2939 (1953).

37. M. M. L. Chen and R. Hoffmann, *J. Am. Chem. Soc.* **98**, 1647 (1976).

38. E. Clementi and D. L. Raimondi, *J. Chem. Phys.* **38**, 2686 (1963).

39. J. W. D. Connolly, in Ref. 186, part A, p. 105.

40. F. A. Cotton, *Group Theory and its Applications,* 2nd ed., Wiley–Interscience, New York (1971).

41. F. A. Cotton and C. S. Kraihanzel, *J. Am. Chem. Soc.* **84**, 4432 (1962).

42. C. A. Coulson, in Ref. 58, pp. 288, 370.

43. L. C. Cusachs, B. L. Trus, D. G. Carroll, and S. P. McGlynn, *Int. J. Quant. Chem.* (Symposium Series) **1**, 423 (1967).

44. J. P. Dahl and C. J. Ballhausen, *Adv. Quantum Chem.* **4**, 170 (1968).

45. B. Davies, A. McNeish, M. Poliakoff, and J. J. Turner, *J. Am. Chem. Soc.* **99**, 7573 (1977).

46. B. M. Deb, *J. Am. Chem. Soc.* **96**, 2030, (1974).

47. M. J. S. Dewar, *Bull. Soc. Chim. Fr.* **18**, C71 (1951).

48. M. J. S. Dewar, *The Molecular Orbital Theory of Organic Chemistry,* McGraw-Hill, New York (1969).

49. M. J. S. Dewar, *Chem. Br.* **11**, 97 (1975).

50. M. J. S. Dewar and R. C. Dougherty, *The PMO Theory of Organic Chemistry,* Plenum, New York (1975).

51. R. S. Drago, *J. Chem. Educ.* **50**, 244 (1973).

52. J. D. Dunitz and L. E. Orgel, *Adv. Inorg. Chem. Radiochem.* **2**, 1 (1960).

53. T. M. Dunn, D. S. McClure, and R. G. Pearson, *Some Aspects of Crystal Field Theory,* Harper and Row, New York (1965).

54. M. Elian and R. Hoffmann, *Inorg. Chem.* **14**, 1058 (1975).

55. P. G. Eller, D. C. Bradley, M. B. Hursthouse, and D. W. Meek, *Coord. Chem. Rev.* **24**, 1 (1977).

56. J. H. Enemark and R. D. Feltham, *Coord. Chem. Rev.* **13**, 339 (1974).

57. S. Esperås, J. W. George, S. Husebye, and Ø. Mikalsen, *Acta Chem. Scand.* **A29**, 141 (1975).

58. H. Eyring, *Physical Chemistry, An Advanced Treatise,* Vol. 5, Academic Press, New York (1970).

59. J. F. Fergusson, *Stereochemistry and Bonding in Inorganic Chemistry,* Prentice-Hall, Englewood Cliffs, N.J. (1974).

60. J. R. Ferraro and G. J. Long, *Acc. Chem. Res.* **8**, 171 (1975).

61. B. N. Figgis, *Introduction to Ligand Fields*, Wiley, New York (1966).

62. I. Fleming, *Frontier Orbitals and Organic Chemical Reactions*, Wiley, New York (1976).

63. K. F. Freed, in Ref. 193, part A, p. 201.

64. K. F. Freed, *Chem. Phys. Lett.* **2**, 255 (1968).

65. A. J. Freeman and R. E. Watson, *Phys. Rev.* **120**, 1254 (1960).

66. F. S. Galasso, *Structure and Properties of Inorganic Solids*, Pergamon, New York (1970).

67. R. M. Gavin, *J. Chem. Educ.* **46**, 413 (1969).

68. R. M. Gavin and L. S. Bartell, *J. Chem. Phys.* **48**, 2460, 2466 (1968).

69. P. George and D. S. McClure, *Progr. Inorg. Chem.* **1**, 381 (1959).

70. M. Gerloch and R. C. Slade, *Ligand Field Parameters*, Cambridge University Press, London (1973).

71. R. J. Gillespie, *Molecular Geometry*, Van Nostrand-Rheinhold, London (1972).

72. R. J. Gillespie and R. S. Nyholm, *Q. Rev.* **11**, 339 (1957).

73. B. M. Gimarc, *Molecular Structure and Bonding*, Academic Press, New York (1979).

74. B. M. Gimarc, *J. Am. Chem. Soc.* **92**, 266 (1970); *ibid.* **93**, 593 (1971).

75. N. I. Giricheva, E. Z. Zasorin, G. V. Girichev, K. S. Krasnov, and V. P. Spiridonova, *Zh. Strukt. Khim.* **17**, 797 (1976).

76. J. Glerup, O. Mønsted, and C. E. Schäffer, *Inorg. Chem.* **15**, 1399 (1976).

77. C. Glidewell, *Inorg. Chim. Acta* **12**, 219 (1975).

78. W. A. G. Graham, *Inorg. Chem.* **7**, 315 (1968).

79. H. B. Gray, *J. Chem. Educ.* **41**, 2 (1964).

80. J. S. Griffith, *The Theory of Transition Metal Ions*, Cambridge University Press, London (1961).

81. D. M. Gruen, in *Fused Salts*, B. R. Sundheim, Ed., McGraw-Hill, New York (1964).

82. L. M. Haines and M. B. H. Stiddard, *Ad. Inorg. Chem. Radiochem.* **12**, 53 (1969).

83. M. B. Hall, *J. Am. Chem. Soc.* **100**, 6333 (1978).

84. M. B. Hall, *Inorg. Chem.* **17**, 2261 (1978).

85. V. R. Haegele and D. Babel, *Z. Anorg. Chem.* **409**, 11 (1974).

86. N. B. Hannay, Ed., *Treatise on Solid State Chemistry*, Vols. 1–6, Plenum, New York (1973–76).

87. M. Hargittai and I. Hargittai, *The Molecular Geometries of Coordination Compounds in the Vapour Phase*, Elsevier, Amsterdam (1977).

88. F. R. Hartley, *Chem. Soc. Rev.* **2**, 163 (1973).

89. E. Heilbronner and H. Bock, *The HMO Model and Its Application*, Wiley, New York (1976).

90. G. Herzberg, *Electronic Spectra of Polyatomic Molecules*, Van Nostrand-Rheinhold, New York (1966).

91. P. J. Hay, *J. Am. Chem. Soc.* **100**, 2411 (1978).

92. M. A. Hitchman and J. B. Bremner, *Inorg. Chim. Acta* **27**, L61 (1978).

93. R. Hoffmann, *Acc. Chem. Res.* **4**, 1 (1971).

94. R. Hoffmann, T. A. Albright, and D. L. Thorn, *Pure Appl. Chem.* **50**, 1 (1978).

95. R. Hoffmann, J. M. Howell, and E. L. Muetterties, *J. Am. Chem. Soc.* **94**, 3047 (1972).

96. R. Hoffmann, *J. Chem. Phys.* **39**, 1397 (1963).

97. R. Hoffmann and W. N. Lipscomb, *J. Chem. Phys.* **36**, 2179, 3489 (1962); *ibid.* **37**, 2872 (1962).

98. R. Hoffmann, M. M. L. Chen, M. Elian, A. R. Rossi, and D. M. P. Mingos, *Inorg. Chem.* **13**, 2666 (1974).

99. R. Hoffmann, M. M. L. Chen, and D. L. Thorn, *Inorg. Chem.* **16**, 503 (1977).

100. R. Hoffmann, J. M. Howell, and A. R. Rossi, *J. Am. Chem. Soc.* **98**, 2484 (1976).

101. R. Hoffmann, B. F. Beier, E. L. Muetterties, and A. R. Rossi, *Inorg. Chem.* **16**, 511 (1977).

102. J. E. Huheey, *Inorganic Chemistry*, 2nd ed., Harper and Row, New York (1978).

103. E. J. Jacob and L. S. Bartell, *J. Chem. Phys.* **53**, 2235 (1970).

104. H. A. Jahn and E. Teller, *Proc. Roy. Soc.* **A161**, 220 (1937).

105. D. A. Johnson, *Some Thermodynamic Aspects of Inorganic Chemistry*, Cambridge University Press, London (1968).

106. J. B. Johnson and W. G. Klemperer, *J. Am. Chem. Soc.* **99**, 7132 (1977).

107. K. H. Johnson, *Adv. Quantum Chem.* **7**, 143 (1973).

108. C. K. Jørgensen, *Oxidation Numbers and Oxidation States*, Springer-Verlag, New York (1969).

109. C. K. Jørgensen, *Modern Aspects of Ligand Field Theory*, North-Holland, Amsterdam (1971).

110. C. K. Jørgensen, *Absorption Spectra and Chemical Bonding in Complexes*, Pergamon, New York (1962).

111. C. K. Jørgensen, R. Pappalardo, and H. H. Schmidtke, *J. Chem. Phys.* **39**, 1422 (1963).

112. R. W. Jotham and S. F. A. Kettle, *Inorg. Chim. Acta* **5**, 183 (1971).

113. F. A. Jurnak, D. R. Greig, and K. N. Raymond, *Inorg. Chem.* **14**, 2585 (1975).

114. S. F. A. Kettle, *Theoret. Chim. Acta* **4**, 150 (1966).

115. S. F. A. Kettle, *J. Chem. Soc.* (A) 420 (1966).

116. S. F. A. Kettle, *Coordination Compounds*, Nelson, London (1969).

117. S. F. A. Kettle, *J. Chem. Educ.* **43**, 21, 652 (1966).

118. S. P. Kowalczyk, L. Ley, F. R. McFeely, and D. A. Shirley, *J. Chem. Phys.* **61**, 2850 (1974).

119. G. E. Kimball, *J. Chem. Phys.* **8**, 188 (1940).

120. W. H. Kleiner, *J. Chem. Phys.* **20**, 1784 (1952).

121. W. G. Klemperer, J. K. Krieger, M. D. McCreary, E. L. Muetterties, D. D. Traficante, and G. M. Whitesides *J. Amer. Chem. Soc.* **97**, 7023, (1975).

122. G. Klopman and R. C. Evans, in Ref. 193, part A, p. 29.

123. I. Kohatsu and B. J. Wuensch, *Acta Cryst.* **B27**, 1245 (1971).

124. R. Krishnamurthy and W. B. Schaap, *J. Chem. Educ.* **46**, 799 (1969); *ibid.* **47**, 433 (1970).

125. C. H. Langford and H. B. Gray, *Ligand Substitution Processes*, Benjamin, New York (1965).

126. E. Larsen and G. N. La Mar, *J. Chem. Educ.* **51**, 633 (1974).

127. J. W. Lauher and R. Hoffmann, *J. Am. Chem. Soc.* **98**, 1729 (1976).

128. J. W. Lauher, M. Elian, R. H. Summerville, and R. Hoffmann, *J. Am. Chem. Soc.* **98**, 3219 (1976).

129. S. L. Lawton and R. A. Jacobson, *Inorg. Chem.* **5**, 743 (1966).

130. A. B. P. Lever, *Coord. Chem. Rev.* **3**, 119 (1968).

131. W. N. Lipscomb, in *Boron Hydride Chemistry*, E. L. Muetterties, Ed., Academic Press, New York (1975), p. 39.

132. H. C. Longuet-Higgins and M. de V. Roberts, *Proc. Roy. Soc.* **224**, 336 (1954).

133. H. C. Longuet-Higgins and M. de V. Roberts, *Proc. Roy. Soc.* **230**, 110 (1955).

134. R. L. Martin and A. H. White, *Trans. Metal Chem.* (Monograph Series) **4**, 113 (1968).

135. D. S. McClure, *Advances in the Chemistry Coordination Compounds*, S. Kirschner, Ed., Macmillan, New York (1961), p. 498.

136. S. P. McGlynn, L. G. Vanquickenborne, M. Kinoshita, and D. G. Carroll, *Introduction to Applied Quantum Chemistry*, Holt, Rheinhart and Winston, New York (1972).

137. R. McWeeny, R. Mason, and A. D. C. Towl, *Disc. Faraday Soc.* **47**, 20 (1969).

138. P. de Meester and A. C. Skapski, *J. Chem. Soc.* (Dalton) 424 (1973).

139. D. M. P. Mingos, *Nature* (Phys. Sci.) **236**, 99 (1972).

140. D. M. P. Mingos, *J. Chem. Soc.* (Dalton) 133 (1974).

141. D. M. P. Mingos, *Adv. Organomet. Chem.* **15**, 1 (1977).

142. P. R. Mitchell and R. V. Parish, *J. Chem. Educ.* **46**, 811 (1969).

143. C. E. Moore, *Ionization Potentials and Ionization Limits Derived from the Analyses of Optical Spectra*, NSRDS-NBS 34, National Bureau of Standards, Washington, D.C. (1970).

144. N. F. Mott and I. N. Sneddon, *Wave Mechanics and its Applications*, Dover, New York (1963).

145. E. L. Muetterties and L. J. Guggenberger, *J. Am. Chem. Soc.* **96**, 1748 (1974).

146. E. L. Muetterties and R. A. Schunn, *Q. Rev.* **20**, 245 (1966).

147. K. W. Muir, in Ref. 193, Vol. 1 (1973), p. 631.

148. R. S. Mulliken, C. A. Rieke, D. Orloff, and H. Orloff, *J. Chem. Phys.* **17**, 1248 (1949).

149. R. S. Mulliken and W. C. Ermler, *Diatomic Molecules*, Academic Press, New York (1977).

150. J. N. Murrell and A. J. Harget, *Semi-Empirical Self-Consistent-Field Molecular Orbital Theory of Molecules*, Wiley, New York (1972).

151. J. N. Murrell, S. F. A. Kettle, and J. M. Tedder, *The Chemical Bond*, Wiley, New York (1978).

152. J. I. Musher, *J. Am. Chem. Soc.* **94**, 1370 (1972).

153. J. I. Musher, *Angew. Chem.* (Int. Ed.) **8**, 54 (1969).

154. W. Nowacki, *Schweiz. Mineral. Petrol. Mitteil.* **49**, 109 (1969).

155. M. O'Keeffe and B. G. Hyde, *Acta Crystallogr.* **B34**, 3519 (1978).

156. L. E. Orgel, *An Introduction to Transition Metal Chemistry*, 2nd ed., Wiley, New York (1966).

157. P. L. Orioli, *Coord. Chem. Rev.* **6**, 285 (1971).

158. E. Parthé, *Crystal Chemistry of Tetrahedral Structures*, Gordon and Breach, New York (1964).

159.   L. Pauling, *The Nature of the Chemical Bond*, Cornell University Press, Ithaca, N.Y. (1960).

160.   R. G. Pearson, *Proc. Natl. Acad. Sci. USA* **72**, 2104 (1975).

161.   R. G. Pearson, *J. Chem. Phys.* **52**, 2167 (1970); *ibid.* **53**, 2986 (1970).

162.   R. G. Pearson, *J. Am. Chem. Soc.* **91**, 1252, 4947 (1969).

163.   W. B. Pearson, *The Crystal Chemistry and Physics of Metals and Alloys*, Wiley, New York (1972).

164.   W. B. Pearson, *J. Phys. Chem. Solids* **23**, 103 (1962).

165.   W. B. Pearson, in Ref. 86, Vol. I, p. 115.

166.   J. R. Perumareddi, *Coord. Chem. Rev.* **4**, 73 (1969).

167.   M. F. Perutz, *Nature* **228**, 726 (1970).

168.   S. D. Peyerimhoff, R. J. Buenker, and L. C. Allen, *J. Chem. Phys.* **45**, 734 (1966).

169.   C. S. G. Phillips and R. J. P. Williams, *Inorganic Chemistry*, Oxford University Press, London (1965, 1966).

170.   J. C. Phillips, in Ref. 86, Vol. I, page 1.

171.   A. Pidcock, R. E. Richards, and L. M. Venanzi, *J. Chem. Soc.* (A) 1707 (1966).

172.   G. C. Pimentel, *J. Chem. Phys.* **19**, 446 (1951).

173.   M. Poliakoff, *Inorg. Chem.* **15**, 2022, 2892 (1976).

174.   J. A. Pople and D. L. Beveridge, *Approximate Molecular Orbital Theory*, McGraw-Hill, New York (1970).

175.   R. D. Poshusta, J. A. Haugen, and D. F. Zetik, *J. Chem. Phys.* **51**, 3343 (1969).

176.   K. F. Purcell and J. Kotz, *Inorganic Chemistry*, Saunders, Philadelphia (1977).

177.   J. W. Richardson, W. C. Nieuwpoort, R. R. Powell, and L. F. Edgell, *J. Chem. Phys.* **36**, 1057 (1962).

178.   N. Rösch and R. Hoffmann, *Inorg. Chem.* **13**, 2656 (1974).

179.   D. R. Rosseinsky and I. A. Dorrity, *Coord. Chem. Rev.* **25**, 31 (1978).

180.   A. R. Rossi and R. Hoffmann, *Inorg. Chem.* **14**, 365 (1975).

181.   R. W. Rudolph, *Acc. Chem. Res.* **9**, 446 (1976).

182.   R. E. Rundle, *Surv. Progr. Chem.* **1**, 81 (1963).

183.   R. E. Rundle, *J. Am. Chem. Soc.* **85**, 112 (1963).

184.   L. Salem, *The Molecular Orbital Theory of Conjugated Systems*, Benjamin, New York (1966).

185.   C. E. Schäffer, *Proc. Roy. Soc.* **A297**, 96 (1967).

186.   C. E. Schäffer, *Wave Mechanics—The First Fifty Years*, S. S. Chissick, W. C. Price, and T. Ravensdale, Eds., Butterworth, London (1973).

187.   C. E. Schäffer, *Structure and Bonding* **14**, 69 (1973).

188.   C. E. Schäffer and C. K. Jørgensen, *Mol. Phys.* **9**, 401 (1965).

189.   C. E. Schäffer, *Pure Appl. Chem.* **24**, 361 (1970).

190.   H. L. Schläfer and G. Gliemann, *Basic Principles of Ligand Field Theory*, Wiley, New York (1969).

191.   G. A. Segal, *Semi-Empirical Methods of Electronic Structure Calculation*, Parts A and B, Plenum, New York (1977).

192.   R. D. Shannon and C. T. Prewitt, *Acta Crystallogr.* **B25**, 925 (1969).

193.   G. A. Sim, L. E. Sutton, Senior Reporters, *Molecular Structure by Diffraction Methods*, Chemical Society Specialist Periodic Reports, London (1973-).

194. O. Sinanoglu and K. Wiberg, Eds., *Sigma MO Theory*, Yale Press, New Haven (1970).

195. J. C. Slater, *Phys. Rev.* **36**, 57 (1930).

196. W. Smith and D. W. Clack, *Rev. Roum. Chim.* **20**, 1243 (1975).

197. R. L. Snow and J. L. Bills, *J. Am. Chem. Soc.* **97**, 6340 (1975).

198. A. Streitweiser, *Molecular Orbital Theory for Organic Chemists*, Wiley, New York (1961).

199. S. Takagi, P. G. Lenhert, and M. D. Joesten, *J. Am. Chem. Soc.* **96**, 6606 (1974).

200. H. B. Thompson and L. S. Bartell, *Trans. Am. Crystallogr. Assoc.* **2**, 190 (1966).

201. J. A. Timney, *Inorg. Chem.* **18**, 2502 (1979).

202. C. A. Tolman, *J. Am. Chem. Soc.* **96**, 2780 (1974).

203. N. Trinajstic, in Ref. 191, Part A, p. 1.

204. A. G. Turner, *Methods in Molecular Orbital Theory*, Prentice Hall, Englewood Cliffs, N.J. (1974).

205. J. H. Van Vleck, *J. Chem. Phys.* **3**, 803, 807 (1935).

206. J. H. Van Vleck, *Phys. Rev.* **41**, 208 (1932).

207. A. Veillard and J. Demuynck, in *Applications of Electronic Structure Theory*, H. F. Schaefer, Ed., Plenum Press, New York (1977), p. 187.

208. L. M. Venanzi, *Chem. Br.* **4**, 162 (1968).

209. T. C. Waddington, *Adv. Inorg. Chem. Radiochem.* **1**, 157 (1959).

210. K. Wade, *Chem. Comm.* 792 (1971).

211. K. Wade, *Chem. Br.* **11**, 177 (1975).

212. K. Wade, *Adv. Inorg. Chem. Radiochem.* **18**, 1 (1976).

213. A. D. Walsh, *J. Chem. Soc.* 2260, 2266, 2288, 2296, 2301, 2306 (1953).

214. K. D. Warren, *Inorg. Chem.* **16**, 2008 (1977).

215. M. Webster and P. H. Collins, *J. Chem. Soc.* (Dalton) 588 (1973).

216. M. S. Wei, J. H. Current, and J. Gendell, *J. Chem. Phys.* **57**, 2431 (1972).

217. B. Weinstock and G. L. Goodman, *Adv. Chem. Phys.* **9**, 169 (1966).

218. A. F. Wells, *Structural Inorganic Chemistry*, 4th ed., Oxford University Press, New York, (1975).

219. M.-H. Whangbo and R. Hoffmann, *J. Chem. Phys.* **68**, 5498 (1978).

220. G. W. Wheland and D. E. Mann, *J. Chem. Phys.* **17**, 264 (1949).

221. R. E. Williams, *Adv. Inorg. Chem. Radiochem.* **18**, 67 (1976).

222. S. Wolfe, *Acc. Chem. Res.* **5**, 102 (1972).

223. M. Wolfsberg and L. Helmholz, *J. Chem. Phys.* **20**, 837 (1952).

224. Y. W. Yared, S. L. Miles, R. Bau, and C. A. Reed, *J. Am. Chem. Soc.* **99**, 7076 (1977).

225. J. J. Zuckerman, *J. Chem. Educ.* **42**, 315 (1965).

Index

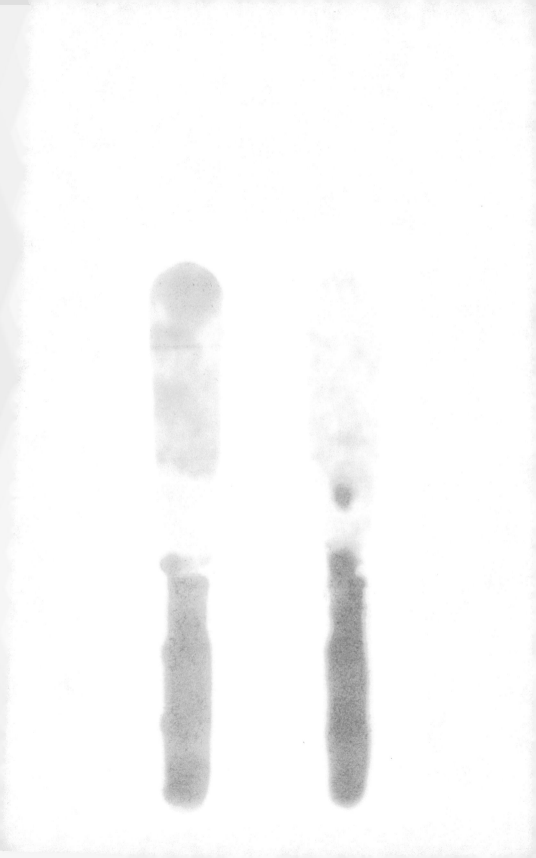